**NANOTECHNOLOGY SCIENCE AND TECHNOLOGY**

# LOW-K NANOPOROUS INTERDIELECTRICS: MATERIALS, THIN FILM FABRICATIONS, STRUCTURES AND PROPERTIES

# NANOTECHNOLOGY SCIENCE AND TECHNOLOGY

Additional books in this series can be found on Nova's website at:

https://www.novapublishers.com/catalog/index.php?cPath=23_29&seriesp=Nanotechnology+Science+and+Technology

Additional E-books in this series can be found on Nova's website at:

https://www.novapublishers.com/catalog/index.php?cPath=23_29&seriesp=Nanotechnology+Science+and+Technology

NANOTECHNOLOGY SCIENCE AND TECHNOLOGY

# LOW-K NANOPOROUS INTERDIELECTRICS: MATERIALS, THIN FILM FABRICATIONS, STRUCTURES AND PROPERTIES

**MOONHOR REE
JINHWAN YOON
AND
KYUYOUNG HEO**

Novinka
Nova Science Publishers, Inc.
*New York*

Copyright © 2010 by Nova Science Publishers, Inc.

**All rights reserved.** No part of this book may be reproduced, stored in a retrieval system or transmitted in any form or by any means: electronic, electrostatic, magnetic, tape, mechanical photocopying, recording or otherwise without the written permission of the Publisher.

For permission to use material from this book please contact us:
Telephone 631-231-7269; Fax 631-231-8175
Web Site: http://www.novapublishers.com

### NOTICE TO THE READER

The Publisher has taken reasonable care in the preparation of this book, but makes no expressed or implied warranty of any kind and assumes no responsibility for any errors or omissions. No liability is assumed for incidental or consequential damages in connection with or arising out of information contained in this book. The Publisher shall not be liable for any special, consequential, or exemplary damages resulting, in whole or in part, from the readers' use of, or reliance upon, this material.

Independent verification should be sought for any data, advice or recommendations contained in this book. In addition, no responsibility is assumed by the publisher for any injury and/or damage to persons or property arising from any methods, products, instructions, ideas or otherwise contained in this publication.

This publication is designed to provide accurate and authoritative information with regard to the subject matter covered herein. It is sold with the clear understanding that the Publisher is not engaged in rendering legal or any other professional services. If legal or any other expert assistance is required, the services of a competent person should be sought. FROM A DECLARATION OF PARTICIPANTS JOINTLY ADOPTED BY A COMMITTEE OF THE AMERICAN BAR ASSOCIATION AND A COMMITTEE OF PUBLISHERS.

**LIBRARY OF CONGRESS CATALOGING-IN-PUBLICATION DATA**

Available upon Request
ISBN: 978-1-61668-749-6

*Published by Nova Science Publishers, Inc.* ✢ *New York*

# CONTENTS

| | | |
|---|---|---|
| **Preface** | | vii |
| **Chapter I** | Introduction | 1 |
| **Chapter II** | Recent Developments in Low-K Nanoporous Dielectrics | 3 |
| **Chapter III** | Characterization of Pore Structures | 29 |
| **Chapter IV** | Conclusions | 47 |
| **References** | | 51 |
| **Index** | | 63 |

# PREFACE

The use of low dielectric constant (low-$k$) interdielectrics in multilevel structure integrated circuits (ICs) can lower line-to-line noise in interconnects and alleviate power dissipation issues by reducing the capacitance between the interconnect conductor lines. Because of these merits, low-$k$ interdielectric materials are currently in high demand in the development of advanced ICs. One important approach to obtaining low-$k$ values is the incorporation of nanopores into dielectrics. The development of advanced ICs requires a method for producing low-$k$ dielectric materials with uniform distributions of unconnected, closed, individual pores with dimensions considerably smaller than the circuit feature size. Thus the control of both pore size and pore size distribution is crucial to the development of nanoporous low-$k$ dielectrics. This article reviews recent developments in the imprinting of closed nanopores into spin-on materials to produce low-$k$ nanoporous interdielectrics for the production of advanced ICs. This review further provides an overview of the methodologies and characterization techniques used for investigating low-$k$ nanoporous interdielectrics.

*Chapter I*

# INTRODUCTION

Continuous improvements in device density and performance have been achieved through feature size reduction and the scaling down of device dimensions to the deep submicrometer level. The coupling of the intermetal capacitance effect with line resistivity is now a limiting factor for the ultra-large-scale integration of electric circuits. To reduce this problem, low-*k* interdielectrics have received significant attention from the microelectronics industry and end users because their use in integrated circuits (ICs) with multilayer structures can lower line-to-line noise in interconnects and alleviate power dissipation issues by reducing the capacitance between the interconnect conductor lines.[1-8] Further, low-*k* interdielectrics have advantages over low-resistivity metal conductors such as copper and silver, because in addition to providing device speed improvements they also provide lower resistance-capacitance delay.[1-7, 9, 10] Thus there is a strong demand for such materials ($k \ll 2.5$) in the microelectronics industry, which is rapidly developing advanced ICs with multilayer structures that have improved functionality and speed in a smaller package and that consume less power.[1-10]

According to the Semiconductor Industry Association's *International Technology Roadmap for Semiconductors*, materials that deliver an effective *k* of 2.5–3.0 are in production today, and that in the near future material systems that deliver an effective $k < 1.9$ are expected to be available, in particular for 50 nm or less feature size technology based on copper metallization.[10] Apart from having a low *k* value, interdielectric materials must meet the thermal and mechanical stabilities requirements of the metallization processing of ICs. For example, copper metallization can be achieved by electroplating, electroless plating, plasma vapor deposition or chemical vapor deposition.[2, 9, 11] These processes are conducted at temperatures below 250 °C but are usually followed by thermal annealing in the range 400–450 °C to ensure the production of void-free copper deposits. Thus, low-$k$ nanoporous materials must be able to withstand thermal stress for several hours. Further, interdielectric materials must have low moisture uptake, high purity, good adhesion to silicon substrate, silicon oxide, and metals, good planarization behavior, and appropriate plasma etching behavior. When copper began to be used as an interconnect metal, damascene (metal inlay) metallization was introduced because of copper's slow dry etching and gas phase deposition processability.[2, 6, 7, 9, 11] The 'dielectric first' damascene process is usually preferred, with trenches then filled with the interconnect metal. In this process, excess metal is removed by chemical mechanical polishing (CMP). The CMP process is conducted with an aqueous slurry containing an abrasive (e.g., alumina particles) and an oxidant and/or complexing agent (e.g., nitric acid or ammonium hydroxide). Therefore, interdielectric materials must be able to withstand harsh CMP processing. Further, diffusion barriers for copper and adhesion promoting layers are necessary.

To achieve these requirements to use for interdielectric materials, much effort has been directed towards the development of low-$k$ porous dielectric thin films for use in the advanced ICs. This article explores recent developments in the imprinting of closed nanopores into spin-on-dielectrics to produce low-$k$ nanoporous interdielectric thin films. We also provide an overview of the analytical techniques used to characterize the pore structures of nanoporous dielectric thin films, and identify the strengths and weaknesses of these techniques. In particular, we discuss in detail the advanced grazing incidence X-ray scattering (GIXS) technique, which has recently gained considerable attention.

*Chapter II*

# RECENT DEVELOPMENTS IN LOW-*K* NANOPOROUS DIELECTRICS

For twenty years, significant effort with polymeric system has been applied to the development of low-$k$ interdielectrics for use in the development of advanced ICs For instance, polyimides, heteroaromatic polymers, polyaryl ethers, fluoropolymers, nonpolar hydrocarbon polymers, and polysilsesquioxanes have been used to low dielectric materials thin films, which deposited from the gas phase with chemical deposition, plasma enhanced chemical vapor deposition, and other techniques.[8, 11-42] However, most of these polymers have $k > 2.5$. Polytetrafluoroethylene (PTFE: Teflon™) has a $k$ value of 2.2, which is the lowest $k$ value reported so far for such polymers. However, PTFE cannot be used in the fabrication of ICs because of its very weak mechanical properties, poor interfacial adhesion and poor processability. Overall, these polymers' $k$ values are considerably lower than those of today's workhorse dielectrics silicon dioxide ($k$ = 3.9–4.3) and silicon nitride ($k$ = 6.0–7.0), but are still much higher than that of air (or vacuum), $k$ = 1.01, which is the lowest value attainable. Hence there has been much interest in incorporating air into dielectric materials as nanopores to produce nanoporous materials with low $k$ values.[1-8] The advanced ICs requires nanoporous materials with a uniform distribution of closed pores that have dimension significantly smaller than a circuit feature size in order to avoid circuit defects.[2-4, 6, 7, 11, 43-45]

In this chapter, we provide an overview of recent developments in the imprinting of closed nanopores into spin-on-dielectrics to produce low-$k$ nanoporous interdielectric thin films.

## II-1. HOLLOW NANOPARTICLES

Thermally and dimensionally stable hollow nanoparticles have been used to produce low-$k$ dielectric materials.[1] For example, films of poly($p$-phenylene biphenyltetracarboximide) (BPDA-PDA PI) containing silica hollow sphere particles have been reported to be useful as low-$k$ polymer dielectric materials.[1] BPDA-PDA PI films containing 27 wt% hollow silica nanoparticles have a refractive index $n$ of 1.7007 at a wavelength of 830 nm, which is less than that of BPDA-PDA PI films ($n$=1.7421). This was the first report of the successful incorporation of closed air voids into dielectric materials for the production of low-$k$ dielectrics. The hollow nanoparticles used in this study were prepared by the thermal sintering of monolithic silica aerogels; the monolithic aerogels were fabricated via the conventional sol-gel reactions of tetraalkoxysilane (TAOS) and its derivatives and subsequent drying under supercritical conditions with selected solvent systems (Figure 1).[46-48] These hollow nanoparticles are based on networked silica and are thus very stable thermally and dimensionally up to 500°C; these properties make them suitable for incorporation into dielectric materials as closed nanopores. Because of their high thermal and dimensional stability, this type of hollow nanoparticle can be blended into any polymer dielectric, including organic polymers such as polyimides and other high temperature polymer dielectrics and inorganic dielectrics such as silicates and organosilicates. If such hollow nanosphere particles were not dimensionally stable because of their chemical and morphological nature, they could easily collapse due to capillary pressure or the molecular mobility that arises in the thermal cycles used in the IC fabrication process.

As described above, silica hollow sphere particles can be used to produce low-$k$ dielectric materials. However, some critical issues for this approach remain to be solved, as follows. Firstly, the silica hollow sphere particles have diameters of approx. 150 nm. The resulting nanopores are thus too large when compared to the metal feature sizes of advanced ICs.

Thus a preparation method must be developed to reduce their size below 10 nm (ideally 5 nm or less). Secondly, the hollow nanoparticles should be well dispersed in the dielectric matrix to produce a high quality interdielectric thin film layer appropriate for the multilayer structure build-up process and with the required dielectric and mechanical properties. Finally, good interfacial adhesion between the hollow nanoparticles and the dielectric matrix is required to achieve the required mechanical properties, which are directly related to the reliability of IC products.

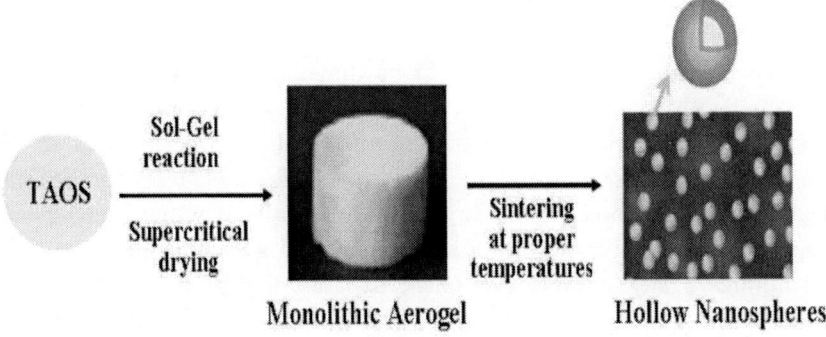

Figure 1. Procedure for preparation of hollow silicate nanoparticles from tetraalkoxysilane (TAOS).

## II-2. DENDRIMERS

Dendrimers possess three distinguishing architectural components—an initiator core, interior layers (the so-called 'generations'), and terminal end groups,[3, 49-55]—and consist of a well-defined, highly branched, compartmentalized structure that is spherical in shape and of nanometer scale.3, 49-55 Dendrimers exhibit unique properties, such as good solubility, low viscosity, multivalence, and encapsulation effects, which result mainly from their branching and spherical architectures.[3, 49-55] A number of dendrimers have been reported in the literature: aliphatic poly(amidoamine), poly(propylene imine), and polyester dendrimers; aromatic polyether, polyester, polyamide, polyimide, polysulfide, and polysulfone dendrimers. [3, 49-55]

Most aliphatic dendrimers have limited thermal stability, i.e. they are thermally degraded below 400°C even in a nitrogen atmosphere, because of their chemical nature.[3, 49-53] Because of this limited thermal stability, their spherical molecular shape, and their nanoscale size, they are suitable for use as thermally labile porogens (i.e., pore generators) for imprinting closed nanopores in dielectric materials through their sacrificial thermal degradation.

However, both dendrimer porogens and dielectric materials must meet the following requirements if they are to be used to successfully fabricate low-$k$ dielectrics containing closed nanopores. Firstly, the dendrimer porogen should thermally degrade at temperatures lower than the degradation temperature of the dielectric material. Secondly, the dielectric material component must be dimensionally stable or become dimensionally stable during the thermal processes required to burn out the dendrimer porogen component from the dielectric film and in the fabrication of ICs. Thirdly, the dendrimer and dielectric components should homogeneously dissolve in a mutual solvent without any phase separation. Fourthly, the components must be highly miscible to prevent or minimize any unfavorable phase separation during film formation processing, i.e., solution casting and subsequent drying processes. Finally, both the dendrimer and dielectric components must retain their miscible state without any unfavorable phase separation until the dendrimer porogen is thermally burned out during the post-thermal processing of the dried film, at which point the imprints of the dendrimer molecules are created as nanopores in the resulting dielectric film.

Polyalkylsilsesquioxanes (PASSQ: $(RSiO_{3/2})_n$, R is an alkyl group) are good dielectric candidates because of their relatively low $k$ values (2.6–3.2), minimal moisture uptake, and high thermal and dimensional stability.[40-42, 56, 57] A variety of soluble PASSQ precursors and their copolymers with a weight average molecular weight $\overline{M}_w$ of less than 20,000 have been reported. These dielectric precursors are curable, so become thermally and dimensionally stable network dielectrics as a result of chemical or thermal treatments. In the case of thermal treatment, these precursors are known to undergo curing reactions (i.e., secondary polycondensation) in the range 75–340°C. A solution of the curable PASSQ precursor and the thermally labile dendrimer porogen in a mutual solvent is spin-coated and cured; a porous structure is then generated by the sacrificial thermal decomposition of the porogen molecules. Removal of the porogen below 400°C yields the desired nanoporous organosilicate; the pore size depends on the size of the dendrimer porogen as well as on the degree of porogen aggregation.

A good example of a dendrimer porogen is shown in Figure 2. Globular poly(propylenimine dotriacontaamine) with 64 ethyl acrylate terminal groups (EA-PPI-64) and poly(propylenimine tetrahexacontaamine) with 128 ethyl acrylate terminal groups (EA-PPI-128) were found to act as good thermally labile porogens in a curable polymethylsilsesquioxane (PMSSQ) dielectric precursor.[3] These dendrimers were found to be miscible with the PMSSQ precursor, and their sacrificial thermal decompositions result in closed, spherical nanopores in the cured PMSSQ dielectric thin films (Figure 3). Grazing incidence X-ray scattering (GIXS) measurements found that loadings in the range 10–40 wt% of the EA-PPI-64 porogen imprint nanopores in the PMSSQ film with an average radius of gyration, $\overline{R}_g$, of 1.4–3.0 nm, which corresponds to an average radius, $\overline{r}$, of 1.1–1.4 nm; the nanopores (1.4–1.5 nm $\overline{R}_g$) imprinted with 10–20 wt% porogen are comparable in size to a single porogen molecule (1.4 nm $\overline{R}_g$), while those (2.0–3.0 nm $\overline{R}_g$) imprinted with 30–40 wt% porogen are slightly larger than a single porogen molecule (Figure 4a). In contrast, the EA-PPI-128 porogen was found to imprint nanopores with an $\overline{R}_g$ of 1.6–1.7 nm, (1.5–1.6 nm $\overline{r}$) in the dielectric films, which are comparable in size to single porogen molecules (1.6 nm $\overline{R}_g$), for porogen loadings in the range 10–40 wt% (Figure 4b).

These results indicate that the EA-PPI-128 porogen, which is filled more densely with its 128 end-groups, is well dispersed at the molecular level in the PMSSQ matrix and can be used to successfully imprint pores with the same size as individual porogen molecules. The EA-PPI-64 porogen has some tendency to aggregate in the PMSSQ matrix, but only very weakly even in the case of a high loading of 40 wt%. Such aggregation may occur as a result of the limited impingement of the end-groups, which have excessive space around them due to the lower number of end-groups in this porogen.

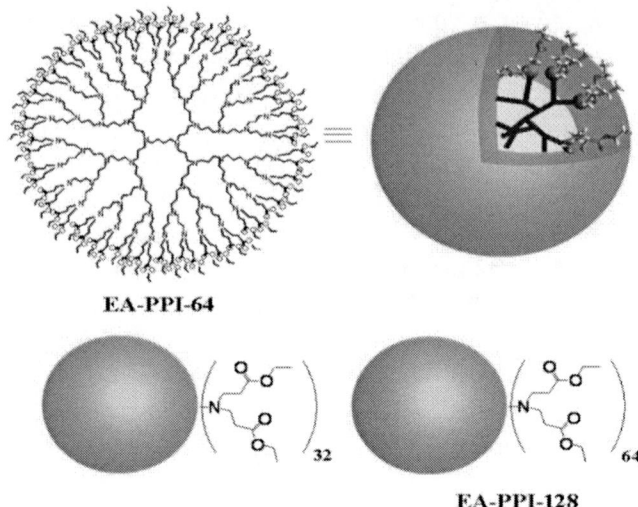

Figure 2. Thermally labile globular dendrimer porogens: EA-PPI-64, ethyl acrylate terminated poly(propylenimine dotriacontaamine); EA-PPI-128, ethyl acrylate terminated poly(propylenimine tetrahexacontaamine).

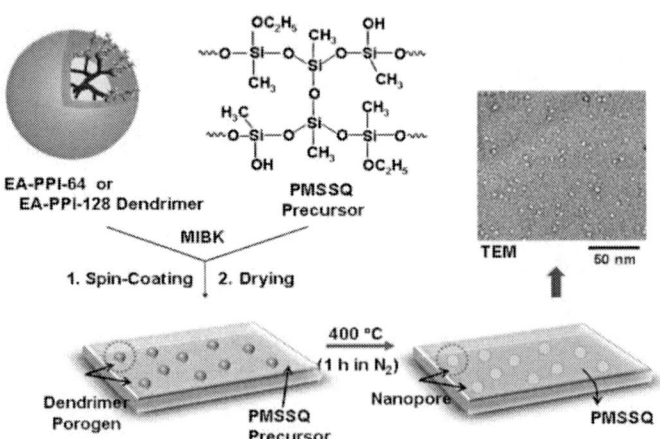

Figure 3. Procedure for preparation of a nanoporous organosilicate dielectric thin film from a curable polymethylsilsesquioxane (PMSSQ) precursor matrix and a thermally labile globular dendrimer porogen (EA-PPI-64 or EA-PPI-128). Transmission electron microscopy (TEM) image of a nanoporous PMSSQ dielectric prepared from a PMSSQ precursor sample loaded with 10.0 wt% EA-PPI-128 porogen.

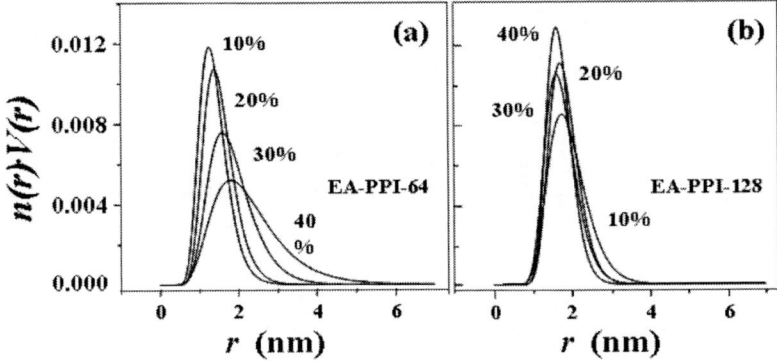

Figure 4. Pore radius and distribution determined from the grazing incidence X-ray scattering (GIXS) analysis: (a) porous PMSSQ films imprinted with EA-PPI-64 porogen and (d) porous PMSSQ films imprinted with EA-PPI-128 porogen. The percentages indicate the initial porogen loadings.

The resulting nanoporous PMSSQ films have porosities in the range 8.6–37.2% depending on the porogens and their loadings as well as the method of determination. For initial porogen loadings up to 40 wt%, the nanoporous films' $k$ values are in the range 1.66–1.71, well below that of the PMSSQ film ($k = 2.70$), and the refractive indices $n$ at a wavelength of 633 nm are in the range 1.253 to 1.260, and thus less than that of the PMSSQ film ($n = 1.396$).

The ethyl acrylate terminated polypropylenimine dendrimers thus exhibit excellent performance as porogens, which derives from their good miscibility with the PMSSQ precursor. However, amine-terminated dendrimers are immiscible with the PMSSQ precursor even in good solvents, and further accelerate the secondary polycondensation of the PMSSQ precursor due to the catalytic activity of their amino end groups, with the undesirable outcome that the resulting PMSSQ product precipitates from solution.[3] Other terminal groups such as 1,4-epoxybutane and butyl glycidyl ether are found to cause severe porogen aggregation, which produces large pores in the resulting dielectric films.[3] These results indicate that the performance of a dendrimer as a labile porogen is strongly dependent upon its terminal groups, which play a major role in determining its miscibility with dielectric materials. Moreover, a molecular level understanding of the size and number of terminal groups of the dendrimer porogen, as well as of the terminal groups' chemical interactions with the dielectric material, is essential for the imprinting of small and well-dispersed closed pores into dielectric materials for use in low-$k$ dielectrics.

## II-3. STAR-SHAPE POLYMERS

Star-shape polymers are structures in which all the chains are linked to a small-molar-mass core.[58, 59] Generally, star-shape polymers have smaller hydrodynamic dimensions than linear polymers with identical molar mass.[60-64] Interest in star-shape polymers arises not only from their use as models for branched polymers but also because of their enhanced segment densities.[60-64]

Star-shape polymers, in particular aliphatic star-shape polymers, are very attractive as porogens for imprinting closed nanopores in dielectrics because of their spherical shape in the nanometer size range. Some aliphatic star-shape polymers that completely decompose at temperatures ≤ 400°C even in an inert atmosphere have been reported, such as star-shape poly(ε-caprolactone)s (PCLs)[4, 5, 43, 58, 65-70] and poly(methyl methacrylate) (PMMA) derivatives.[71]

Among the star-shape polymers reported,[43, 58-71] star-shape PCLs have been extensively investigated for use as porogens in PASSQ dielectrics because of the following chemical characteristics.[4, 5, 43, 58, 65-70] PCL polymers consist of nonpolar pentylenyl and polar ester segments in each repeat unit in the backbone. In addition, they have hydroxyl groups at their arm ends. The PMSSQ dielectric precursor (i.e., a PASSQ precursor) contains hydroxysilyl and alkoxysilyl groups. The PCL polymers are therefore likely to be miscible with the PMSSQ precursor. Star-shape PCLs with 4–48 arms have been reported.[4, 5, 43, 58, 65-70]

Loadings of star-shape PCL porogens with 4 and 6 arms (PCL4 and PCL6) (Figure 5) of up to 20 wt% into the PMSSQ precursor produce optically clear blend films, and result in pores with $\overline{R}_g$ ranging from 5.3 to 14.2 nm in the cured dielectric films after sacrificial thermal degradation.[4, 5, 72] For initial porogen loadings up to 20 wt%, the films' $k$ values were found to be in the range 1.90–2.16, down from $k =$

2.70, and the refractive indices $n$ at a wavelength of 633 nm were found to be in the range 1.2904–1.3207, down from 1.396.

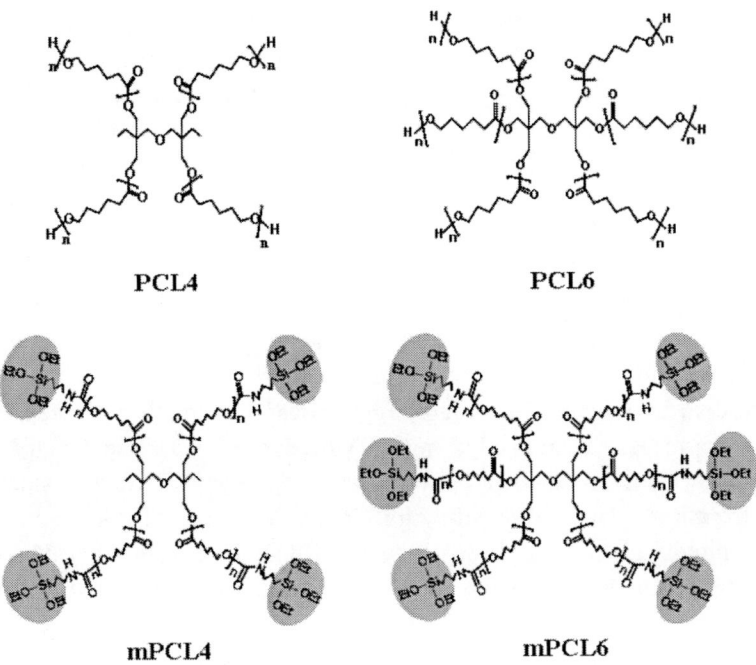

Figure 5. Star-shape polymer porogens: PCL4, four-armed poly(ε-caprolactone) (PCL); mPCL4, triethoxysilyl-terminated four-armed PCL; PCL6, six-armed poly(ε-caprolactone) (PCL); mPCL6, triethoxysilyl-terminated six-armed PCL.

The resulting nanopores are larger than single porogen molecules, indicating that aggregation of the porogens occurs even at loadings ≤20 wt%. At porogen loadings ≥30 wt%, significant phase separation occurs, generating large and interconnected pores in the resulting delectric films.[4, 5, 43, 65-70, 72] Furthermore, as the number of arms in the porogen increases, severe porogen aggregation occurs even at porogen loadings as low as 10 wt%, producing very large and interconnected pores in the dielectric films.[43]

The aggregation behavior of PCL4 porogen was recently investigated in detail by carrying out an in-situ GIXS analysis with synchrotron X-ray radiation sources to determine the mechanism of nanopore formation.[5] This study obtained the following results. The porogen and precursor polymer are miscible in the blend films. However, during heat treatment, aggregation of the porogen molecules is induced below 200 °C by the curing reaction of the PMSSQ precursor matrix. This porogen aggregation is attributed to several principal factors. Firstly, curing results in the formation of PMSSQ precursor crosslinks, and thus in the segregation of the precursor molecules. Secondly, curing of the precursor molecules also produces ethanol and water byproducts, which are removed. Byproduct formation and removal convert the polar PMSSQ precursor molecules into the nonpolar crosslinked PMSSQ dielectric. Finally, because the porogen molecule has only four arms, there is excessive space around each arm, which permits the approach of the arms of other porogen molecules, leading to their segregation and aggregation. During the heat treatment process these factors all contribute to the generation of porogen aggregates; the shape, size, and size distribution of the porogen aggregates are directly reflected in the dimensions of the imprinted pores. Moreover, it was found that higher porogen loadings result in larger porogen aggregates with a broader size distribution. Thus the structural characteristics of the nanopores imprinted within the PMSSQ dielectric films are governed by the nature of the PCL4 porogen aggregates formed during curing of the PMSSQ precursor matrix.

Thus star-shape PCL porogens have a tendency to aggregate that is worsened by the crosslinking of the PMSSQ precursor matrix. Other star-shape porogens exhibit similar problems.[71]

As discussed above, star-shape aliphatic polymers are candidates for use as porogens because of their spherical shape and nanoscale size, but their tendency to aggregate limits their ability to create small pores and enhance the porosity of the resulting dielectrics,[4, 5, 43, 65-71, 73] making them unsuitable for use in advanced ICs patterned with small feature sizes. Therefore, the challenge remains to prevent or minimize the aggregation of star-shape polymer porogens in the dielectric matrix throughout the dielectric film formation process.

One approach attempts to minimize severe aggregation of star-shape polymer porogens through the chemical modification of the porogen end-groups (Figure 5).[4, 73-75] For example, the hydroxyl ends of star-shape PCL porogen can be modified with triethoxysilyl groups, which are analogs of the reactive functional groups of the PMSSQ precursor that take part in curing reactions during thermal treatment. GIXS analysis found that triethoxysilyl-terminated PCL4 (mPCL4) can be used to imprint nanopores with $\overline{R}_g$ of 5.2–17.1 nm in PMSSQ dielectric films for 10–40 wt% loadings.[73, 74] The nanopores (5.2–10.0 nm $\overline{R}_g$) imprinted with 10–20 wt% mPCL4 are comparable in size to those created with 10–20 wt% loadings of the corresponding PCL4 porogen; however, the pores (15.1–17.1 nm $\overline{R}_g$) imprinted with 30–40 wt% mPCL4 are much smaller than those generated with the same loadings of the corresponding PCL4 porogen. The nanoporous films were found to have porosities in the range 5.8–25.6% for 10–30 wt% porogen loadings. For initial porogen loadings up to 30 wt%, the film's $k$ value is 1.95, down from $k = 2.70$, and the refractive index $n$ at a wavelength of 633 nm is 1.2921, down from 1.396.

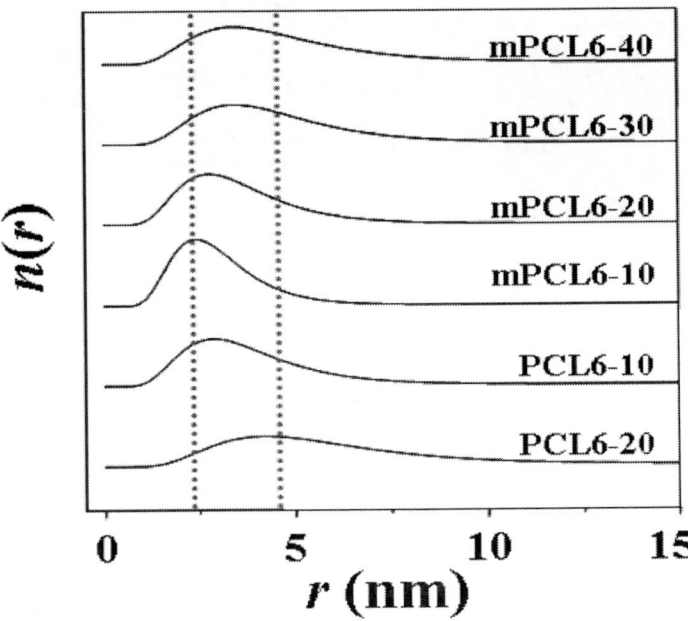

Figure 6. Pore radius and distribution determined from the grazing incidence X-ray scattering (GIXS) analysis: porous PMSSQ films imprinted with PCL6 and mPCL6 porogens. The porous film samples are labelled according to the initial porogen loading in weight per cent: for example, mPCL6-40 correspond to the porous film prepared with 40 wt% mPCL6 porogen.

In the case of triethoxysilyl-terminated PCL6 (mPCL6), the thermal degradation of the porogen was found to imprint nanopores with $\overline{R}_g$ in the range 5.8–12.5 nm in PMSSQ dielectric films, depending on the porogen loading in the range 10–40 wt%; these pores are much smaller than those created with the corresponding PCL6 porogen (Figure 6)[.4, 75] The nanoporous films were found to have porosities in the range 7.8–39.7%, depending on the porogen loading. For initial porogen loadings up to 40 wt%, the films' $k$ value is 1.67, down from $k = 2.70$.

Overall, the use of triethoxysilyl-terminated PCL porogens significantly reduces the aggregation of the porogen molecules for loadings up to 40 wt% in the PMSSQ dielectric throughout the entire film formation process. These results highlight the potential of triethoxysilyl-modification of end groups as a means of preventing severe aggregation of star-shape polymer porogens with large numbers of arms in the preparation of low-$k$ nanoporous organosilicate dielectrics.

## II-4. HYPERBRANCHED POLYMERS

Aliphatic hyperbranched polymers are thermally degradable below 400 °C. They can be synthesized with a molecular size of a few nanometers or less. Further, they have relatively high numbers of end groups, which can favorably interact with dielectric materials such PASSQ precursors. Because of these molecular characteristics, hyperbranched aliphatic polymers have been considered as another porogen candidate for templating nanopores in PASSQ dielectrics.

Hyperbranched poly(ε-caprolactone)s (PCLs) are typical of this class of porogen.[43, 69, 70, 76] Hyperbranched PCLs are more soluble and less crystalline than star-shape PCLs. They have been successfully loaded at concentrations up to 30 wt% into PMSSQ precursor films.[67, 76] Scanning electron microscopy (SEM) was used to show that the resulting pores had a size of 20 nm (i.e., a radius of 10 nm).[76] These results indicate that hyperbranched PCL porogens are more miscible with the PMSSQ precursor matrix than star-shape PCL porogens, and generate smaller pores in the cured dielectric films than star-shape PCLs.[67, 76] The nanoporous films prepared with a porogen loading of 30 wt% were found to have a $k$ value of 2.0.

A block copolymer of hyperbranched PCLs, hyperbranched poly(ε-caprolactone-b-methyl methacrylate), has been described, which can be synthesized by either sequential or concurrent polymerization of γ-(ε-caprolactone)-2-bromo-2-dimethyl-propionate with 2-hydroxyethyl methacrylate in the presence of ε-caprolactone and methyl methacrylate.[77, 78] This hyperbranched block copolymer can be loaded at concentrations up to 30 wt% into PMSSQ precursor films. Transmission electron microscopy (TEM) measurements found that the resulting nanopores had an average size of 8 nm (a radius of 4 nm). This pore size is slightly smaller than that of nanopores imprinted with hyperbranched PCLs. The nanoporous dielectric films imprinted with a porogen loading of 30 wt% were found to have a $k$ value of 2.1.

For these hyperbranched PCLs and their block copolymers, the maximum porogen loading is 30 wt%, i.e., there is a limitation on the loading of these hyperbranched polymer porogens into dielectric films and on the minimum achievable size of the imprinted pores.

Trimethylsilylated hyperbranched polymers have also been used as porogens. They are synthesized via the pseudo one pot polycondensation of 2,2-bis-hydroxymethyl propionic acid with a tetrafunctional ethoxylated pentaerythritol core and subsequent end-group modification with chlorotrimethylsilane (Figure 7).[79] These porogens have been successfully loaded at concentrations up to 40 wt% into PMSSQ precursor films. However, the sizes of imprinted pores range from 15 to 60 nm (i.e., a radius of 7.5–30 nm) depending on the initial porogen loading as well as the number of generations in the hyperbranched polymers. Further, the imprinted pore sizes depend on the degree of trimethylsilylation in the intially loaded porogen; high trimethylsilylation in the porogen significantly reduces the size of the imprinted pores. The porogens with no trimethylsilylation are found to generate very large pores: pores of size >>100 nm are generated depending on the porogen loading.

New hyperbranched polymer porogens based on aliphatic polyethers have recently been prepared and tested as porogens in a copolymer of PMSSQ precursor, poly(methylsilsequioxane-$co$-1,4-bis (ethylsilsesquioxane)-benzene) (PMSSQ-BESSQB), which exhibits better properties than PMSSQ dielectrics.[80] The new porogens are hyperbranched polyglycidol (PG) and its ketalized derivative (K-PG) (Figure 7), which were synthesized via a new polymerization pathway of glycidol. In particular, K-PG has good solubility in common solvents and good miscibility with the PMSSQ-BESSQB precursor. Moreover, K-PG exhibits sacrificial thermal decomposition characteristics that make it suitable for use as a porogen in the fabrication of porous PMSSQ-BESSQB dielectric films. K-PG can be loaded into the PMSSQ-BESSQB precursor at concentrations up to 40 wt%. A GIXS study of the porous thin films prepared from PMSSQ-BESSQB/K-PG composite films with various compositions found that the average size of the pores in the porous dielectric films varies from 6.7 to 18.5 nm (i.e., a radius of 3.4 to 9.3 nm) as the initial loading of the K-PG porogen is increased from 10 to 40 wt%. These pores are spherical and have a sharp interface with the dielectric matrix. As the initial loading of the K-PG porogen is increased up to 40 wt%, the porosities of the PMSSQ-BESSQB films increase almost linearly to 37 vol% and the refractive indices $n$ decrease almost linearly from 1.450 to 1.270. The presence of the imprinted pores reduces the $k$ values of the PMSSQ-BESSQB films almost linearly as the initial loading of the K-PG porogen increases.

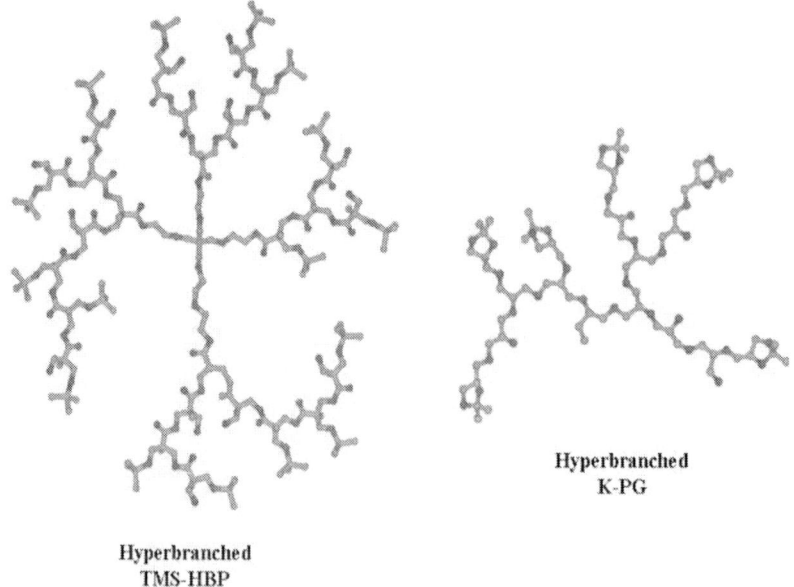

Figure 7. Hyperbranched polymer porogens: TMS-HBP, trimethylsilylated hyperbranched polyester; K-PG, ketalized polyglycidol.

## II-5. CROSSLINKED POLYMER NANOPARTICLES

In recent years nanoparticles have attracted significant attention because of their potential applications in the nanotechnology field.[81-87] In particular, thermally labile organic particles in the nanometer range are very attractive because of their potential use as porogens in the production of low-$k$ nanoporous dielectrics. One approach to fabricating organic nanoparticles is through the self-crosslinking reaction of a crosslinkable polymer.[85-88] This approach relies on the controlled intramolecular crosslinking of a functionalized polymer chain. The shape of the nanoparticle is controlled by the crosslinking chemistry, polymer type, functionality, and architecture.

Examples are crosslinked poly(styrene-co-methacroyloxyethyl methacrylate), poly(ε-caprolactone-co-acryloyloxycaprolactone), and poly(methyl methacrylate-co-methacroyloxyethyl methacrylate) nanoparticles (Figure 8).[88] The syntheses of these crosslinked polymeric nanoparticles consist of two steps. The first step involves the preparation of the potentially crosslinkable macromolecules. In the second step, the nanoparticles are prepared via the self-crosslinking reactions of the crosslinkable macromolecules in ultra-dilute solutions using a radical initiator. The resulting nanoparticles have a hydrodynamic radius of 3.8–13.1 nm; the particle size increases with the molecular weight of the crosslinkable polymer.

It was found that polymer nanoparticles could be used to imprint nanopores in cured PMSSQ dielectric films by their sacrificial degradation through heat treatment up to 450 °C. This approach depends on the solubility of the nanoparticles and the uniform dispersion of the nanoparticles with minimal aggregation in the dielectric matrix. The best results have been achieved with crosslinked poly(methyl methacrylate-co-methacroyloxyethyl methacrylate) nanoparticles. This nanoparticle has a hydrodynamic radius of 6.5 nm and has been found to generate nanopores with a radius of 7.2 nm in cured PMSSQ dielectric films. Nanoporous films imprinted with 20–30 wt% loadings of these nanoparticles were shown to exhibit refractive indices $n$ of 1.31–1.28. The $k$ value of a PMSSQ film with 20% porosity was found to be 2.1, which is a significant reduction below that of the PMSSQ film.

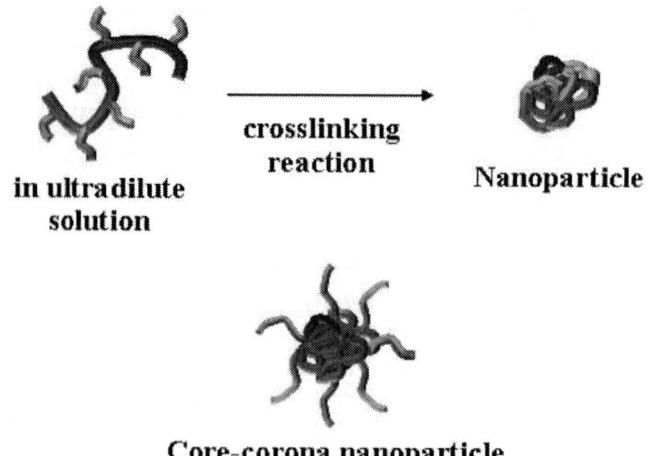

Figure 8. Nanoparticle porogens: a crosslinked polymer nanoparticle and its preparation scheme (e.g., crosslinked poly(ε-carprolactone-co-acryloyloxycarprolactone)) (top); a core-corona nanopoarticle (e.g., a nanoparticle based on a core consisting of polynorbornene blocks and a corona composed of poly(ethylene oxide) blocks) (bottom).

## II-6. CORE-CORONA POLYMER NANOPARTICLES

Core-corona polymers have recently been introduced for use as porogens, particularly norbornene-ethylene oxide copolymers (Figure 8).[89] As shown in Figure 5b, these polymers actually consist of a hyperbranched polymer with norbornene polymer inner parts and ethylene oxide polymer outer parts. The inner parts are insoluble and bulky, and can act a core in solution as well as in a dielectric matrix. The outer parts are polar and soluble, and can act as a corona to favorably interact with the solvent as well as with the dielectric matrix. The outer corona renders the insoluble core compatible with the dielectric matrix and suppresses aggregation and precipitation of the insoluble interior.

Core-corona polymers have been synthesized with diameters of 10–20 nm, depending on the sizes and fractions of the core and corona parts. They are dissolved in a solution with the PMSSQ precursor and the resulting solution is spun onto substrates and thermally treated at 450°C to produce porous PMSSQ dielectric films. A TEM study has found that pores are generated with a size of 10–20 nm in the dielectric films, depending on the porogens and their initial loadings; the imprinted pores are comparable in size to the core-corona polymer nanoparticles. The $k$ value of the resulting dielectric film has been found to decrease from 2.8 to 1.7 with increases in the porogen loading up to 50 wt%.

Other core-corona molecules is octa(2,4-dinitrophenyl)-silssesquioxane (ODNPSQ), which consists of one cubic $Si_8O_{12}$ core covered by eight dinitrophenyl groups as corona. Due to the lipophilic character of 2,4-dinitrophenyl group, this porogen show good miscibility with polyphenylsilsesquioxane (PPSQ), resulting porous spin-on thin film after sacrificial thermal degradation at 450 °C. With the 40% of porogen loading, porous film show low water absorption of 0.45% and low dielectric cnstant of 1.93.[90]

## II-7. LINEAR POLYMERS

Various linear aliphatic homo- and co-polymers have been investigated as porogens for producing nanoporous dielectrics: linear homopolymers,[91, 92] random copolymers,[45, 93-96] amphiphilic diblock copolymers,[97, 98] and triblock copolymers.[99-101]

Representative homopolymer porogens are poly(alkylene ether)s (e.g., poly(ethylene oxide) (PEO) and poly(propylene oxide) (PPO)) and polyesters (e.g., poly(ε-caprolactone) (PCL) and poly(lactic acid) (PLA)).[91, 92] However, these polymers exhibit very limited miscibility with PASSQ precursors, which results in severe phase separation depending on their loading levels, and in large, interconnected pores in cured PASSQ dielectrics.

One example of a random copolymer porogen is poly(methyl methacrylate-co-dimethylaminoethyl methacrylate) (P(MMA-co-DMAEMA)), which was synthesized via the radical copolymerization of methyl methacrylate and $N,N$-dimethylamino ethyl methacrylate.[45, 93, 95] The tertiary amino group in the DMAEMA component of the copolymer favorably interacts with the functional groups (i.e., hydroxysilyl groups) of the PASSQ precursor via strong hydrogen bonding, and thus its presence results good miscibility with the precursor, but it also catalyzes the polycondensation (i.e., sol-gel reaction) of the precursor even at room temperature, causing phase separation and precipitation of the precursor. Because of this dual functionality, the incorporation of the DMAEMA component is restricted to loadings less than 15 mol%, which produces a copolymer miscible with the PASSQ precursor (e.g., PMSSQ precursor) and ultimately generates small pores in cured dielectrics by sacrificial degradation through thermal treatment at 400–450 °C. At a porogen loading of 40 wt%, the $k$ value of the resulting porous dielectric film is decreased to 1.95, which is less than that of PMSSQ dielectrics ($k$ = 2.70).[93] Small angle X-ray scattering (SAXS) and prism coupling measurements on the resulting porous dielectric films showed that for initial porogen loadings in the range 5–50 wt%, the pore sizes range from 2 to 10 nm, the porosities range from 5 to 50 %, and the refractive indices at 633 nm range from 1.35 to 1.21.[45, 93] However, with increases in the porosity, the pore size increases and the pore size distribution broadens, indicating that the generated pores change from closed-cell structures to interconnected bicontinuous structures, which is attributed to changes in the phase separation of the blends of the porogen and matrix components with changes in the blend composition.[45] The pore size and size distribution are also found to be affected by the numbers of hydroxylsilyl and alkoxysilyl groups in the PMSSQ precursor. Moreover, neutron reflectivity measurements on these porous films found localized higher porosities at the interface between the porous films and their silicon substrates.[95]

Poly(styrene-$b$-2-vinylpyridine) (PS-$b$-P2VP) is an example of an amphiphilic diblock copolymer porogen; the P2VP block fraction ranges from 26 to 65 mol%.[98] As for P(MMA-$co$-DMAEMA), PS-$b$-P2VP porogens exhibit good miscibility with PMSSQ precursor via hydrogen bonding interactions between the pyridine rings of the P2VP block in the porogen and the hydroxysilyl groups of the precursor, and can be used to produce small pores in cured dielectric films. As the initial loading of the diblock copolymers is increased up to 60 wt%, the refractive indices $n$ of the resulting porous PMSSQ films decrease almost linearly from 1.361 to 1.139. SAXS analysis found that nanopores with an average size of 11.6 nm were generated in PMSSQ films imprinted with 30 wt% porogen loading.

Poly(ethylene oxide-$b$-propylene oxide-$b$-ethylene oxide) (PEO-$b$-PPO-$b$-PEO) has been tested as an amphiphilic triblock copolymer nanopore template in PMSSQ dielectrics.[99, 101] Positronium annihilation lifetime spectroscopy (PALS) measurements on the resulting porous dielectric films found that closed pores are generated at porogen loadings ≤20 wt% but that interconnected pores are imprinted for porogen loadings >20 wt%. Small angle neutron scattering (SANS) and PALS observations showed that nanopores with sizes in the range 2.2–5.2 nm were generated, depending on the initial porogen loading. The $k$ value was reduced to 1.5 and the porosity increased to 53% as the initial porogen loading was increased to 50 wt%.

Poly(styrene-$b$-3-trimethoxysilylpropyl methacrylate) (PS-$b$-PMSMA: the numbers of repeat units are 118 for the PS block and 12 for the PMSMA block) was synthesized as a reactive linear block copolymer porogen and then tested in a PMSSQ precursor.[97] For initial porogen loadings up to 50 wt%, the PMSSQ film's $k$ value was found to decrease to 1.84, down from $k = 2.70$, and the refractive index $n$ at a wavelength of 633 nm was found to decrease to 1.226, down from 1.354. Atomic force microscopy (AFM) and TEM observations found that for 10–50 wt% porogen loadings, pores 5.2–12.7 nm in size were generated in the dielectric films. These pores are smaller than those (5.4–20.4 nm) imprinted with the same loading of poly(styrene-$b$-acrylic acid) (PS-$b$-PAA: the numbers of repeat units are 30 for the PS block and 58 for the PAA block). However, the reduction in pore size achieved by using this reactive block copolymer is not significant.

## II-8. CAGE SUPRAMOLECULES

Cyclodextrins (CDs) are cyclic oligosaccharides consisting of at least six glucopyranose units joined together by an α-linkage: α-cyclodextrin (α-CD) (6 glucose units), β-cyclodextrin (β-CD) (7 glucose units), and γ-cyclodextrin (γ-CD) (8 glucose units) (Figure 9).[102-105] They are composed of a hydrophobic interior and a hydrophilic exterior; in particular, the hydrophilic exterior may produce favorable interactions with dielectric materials with polar characteristics. These aliphatic compounds are thermally labile cage supramolecules with a maximum diameter of 1.5–2.0 nm. Due to their molecular sizes and characteristics, CDs are potentially useful as porogens. CDs are capable of generating pores in silicate dielectrics prepared via the sol-gel reaction of tetramethoxysilane.[103] However, a TEM study found that these CDs imprinted wormlike pores 1.5 nm in diameter and tens of nanometers long in the silicate dielectrics, which was attributed to their aggregation in a stacking manner in one direction.[103]

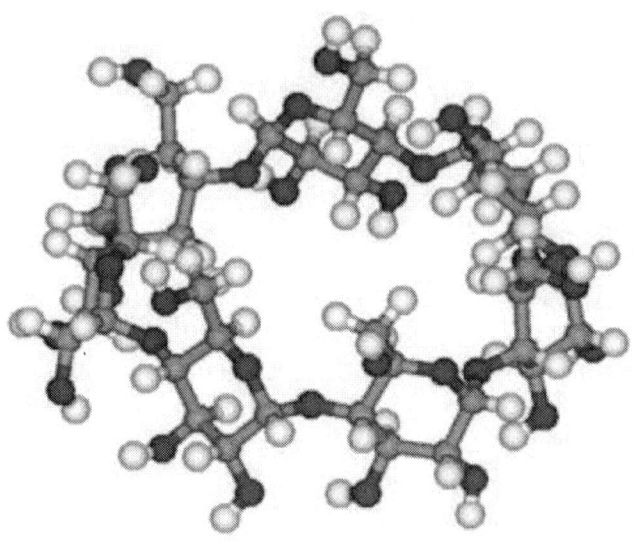

β-CD

Figure 9. β-Cyclodextrin (β-CD), a cage supramolecular porogen.

To improve their miscibility with dielectric materials, the hydroxyl groups of CDs have been modified.[104, 105] Some modified CDs are prepared: methyl-β-CD, ethyl-β-CD, acetyl-β-CD, propanoyl-β-CD, and benzoyl-β-CD.

PALS analysis found that methyl-β-CD can be used to imprint 1.6–2.2 nm nanopores in cured cyclic silsesquioxane (CSSQ) dielectrics by sacrificial thermal degradation at 420 °C, depending on its initial loading in the range 10–40 wt%; the porosities ranged from 9.4 to 25.9%.[104] For initial porogen loadings up to 40 wt%, the nanoporous films' $k$ value was 1.90, down from 2.51, and the refractive index $n$ at a wavelength of 633 nm was 1.335, down from 1.433. As mentioned earlier, PMSSQ dielectric films have $n$ = 1.3936 and $k$ = 2.70. This suggests that all the $k$ values reported for the porous CSSQ films imprinted with methyl-β-CD were under-estimated. However, methyl-β-CD loadings of ≥50 wt% were also found to generate wormlike and interconnected pores, as observed for the porous silicates imprinted with CDs without any modifications. Furthermore, the other modified CDs (ethyl-β-CD, acetyl-β-CD, propanoyl-β-CD, and benzoyl-β-CD) are also found to create such interconnected pores in cured CSSQ and modified CSSQ dielectric films.[104, 105] The lengths of the interconncted pores were found to vary in the range 30–160 nm, depending on the modified functional groups in the β-CD and their loading levels. Thus the affinity between β-CD molecules with different functional groups is crucial to reducing the pore size and pore interconnectivity of porous dielectric films generated with this method.

Furthermore, β-CDs have been modified with chemically reactive triethoxysilyl group. Triehoxysilyl cyclodextrin (TESCD) was synthesized by allylation and hydrosilylation reaction using triethoxysilane. The porosity linearly increased with increase of TESCD loading to poly(methyl trimethoxy silane-co-bistriethoxysilyl ethane) matrix, indicating no pore collapes occurred during pore generation. TESCD porogen resulted in higher mechanical strengths than PCL porogen, and $k$ values reached down to 2.2 at 40% of porogen loading.[106]

## II-9. HIGH BOILING POINT MOLECULES

Soluble PASSQ precursors are known to undergo curing reactions (i.e., secondary polycondensation) in their dried films over the temperature range 75–360 °C in heating runs under a nitrogen atmosphere or in vacuum.[3-5, 73, 74] The curing of PASSQ precursors can even be carried out at room temperature with the aid of catalysts such as primary, secondary, and tertiary amines.[3, 93] Because of these curing characteristics of PASSQ precursors, a solvent with a high boiling point and low vapor pressure can be used to create pores in cured PASSQ dielectric films by carrying out its thermal evaporation after the dielectric precursor film solidifies, as long as the chain extension and crosslinking reactions of the hydroxysilyl and alkoxysilyl groups occur to a certain extent below the boiling point of the solvent.[107, 108]

Several organic solvents with high boiling points and low vapor pressures were tested with the aid of ammonia catalyst as porogens in polyhydrogensilsesquioxane (PHSSQ) dielectrics: laurone, hexadecylhexadecanoate, squalane, didecylphthalate, triaconene, and triaconene.[107, 108] In this approach, the soluble PHSSQ precursor is deposited with a high boiling-point solvent that is insoluble with the PHSSQ precursor solution but compatible with the precursor; the high boiling-point solvent is then present in small domains within the precursor matrix. This deposited film is treated with wet ammonia to create gel state PHSSQ. During curing at 470 °C in a nitrogen atmosphere, the PHSSQ precursor molecules form a network structure and nanopores are generated by boiling out the solvent from the precursor matrix. Average pore sizes range from 2.6 to 5.3 nm depending on which high boiling point solvent is used as the porogen. Among the high boiling point solvents studied so far, laurone yields the smallest pore size, 2.6 nm. A $k$ value of 2.2 was achieved, which is a reduction from that (3.0–3.2) of the PHSSQ dielectric film. However, most of the generated nanopores were found to be open and interconnected rather than closed.

## II-10. HYBRID COPOLYMERS

Hybrid organic-inorganic copolymers composed of thermally labile organic parts that produce pores and thermally curable inorganic or organosilicate precursor parts that become dielectric matrix can be used in another approach to introducing porosity into dielectric materials.[108-110]

A PHSSQ precursor containing $-OC_{20}H_{41}$ groups covalently linked to the precursor backbone was synthesized.[108] Films of this organic-inorganic copolymer were prepared by solution casting and subsequent drying, and then subjected to thermal processing at 450°C in a nitrogen atmosphere, resulting in nanoporous dielectric films.[108] This approach was found to produce dielectric films with 2.2 nm size nanopores and a low $k$ value of 1.8. However, it was confirmed with PALS measurements that interconnected rather than closed pores were formed.

Other hybrid copolymers are poly(methyl-*co*-trifluoropropyl)-silsesquioxanes (P(M-*co*-TFP)SSQs), which can be synthesized from methyltrimethoxysilane and trifluoropropyl-trimethoxysilane.[111] Curing of films of this copolymer at 420 °C produces dielectric films with $k = 2.2$ and $n = 1.348$; these relatively low $k$ and $n$ values are attributed to the porosity generated by the thermal degradation of the trifluoropropyl groups. However, for this dielectric film the reduction in the $k$ value is not significant.

PMSSQ precursors chemically linked with poly(methyl methacrylate) (PMMA) thermally labile branch chains have been described.[112] In these hybrid organic-inorganic block copolymers, the PMSSQ fractions range from 18 to 70 mol%. The weight average molecular weights range from 7,700 to 100,000; hybrid block copolymers with higher PMSSQ fractions can be synthesized with lower molecular weights. These hybrid block copolymers can be used as porogens in cured PMSSQ and its copolymer dielectrics. In the case of a 18 mol% PMSSQ block porogen hybrid, nanopores with a size of 1.7–2.4 nm were generated in cured PMSSQ-based dielectric films by thermal treatment at 430 °C with increases in the initial porogen loading up to 30 wt%. The porous films imprinted with 30 wt% loading of the porogen have a $k$ value of 2.2.

Other hybrid copolymers system prepared with covalently bonded adamantylphenol to PMSSQ. The adamantylphenol groups were grafted or bridged to PMSSQ through propyl linkers and thermal decomposition of such groups occurred through the cleavage of covalent bonds during thermal curing. This organic-inorganic hybrid copolymer was prepared by hydrosilylation between trimethoxylsilane and (allyloxyphenyl) adamantane and then copolymerized with methyltrimethoxysilane (MTMS). Curing of films of this copolymer at 250°C produces dielectric films with $k = 2.3$ and elastic modulus of 5.5 GPa.[110]

Another organic-inorganic hybrid system consists of soluble PMSSQ precursors chemically linked with PCL4 porogens (Figure 10).[109] These hybrid star-block copolymers are synthesized via the sol-gel reactions of triethoxysilyl-terminated PCL4 (mPCL4)[73, 74] and PMSSQ prepolymer in various compositions; in these sol-gel reactions the amount of mPCL4 was adjusted to 10, 20, and 30 wt% relative to the PMSSQ prepolymer, which has a weight-average molecular weight of 3800. Porous PMSSQ dielectric films derived from the hybrid star-block copolymers were prepared with curing at 400°C. In the resulting porous films, nanopores with a size of 5.0 nm were found to be generated. Porous films prepared from a hybrid star-blocked copolymer containing 30 wt% PCL4 segments were found to exhibit a refractive index of 1.320, down from that of PMSSQ dielectrics ($n = 1.396$).

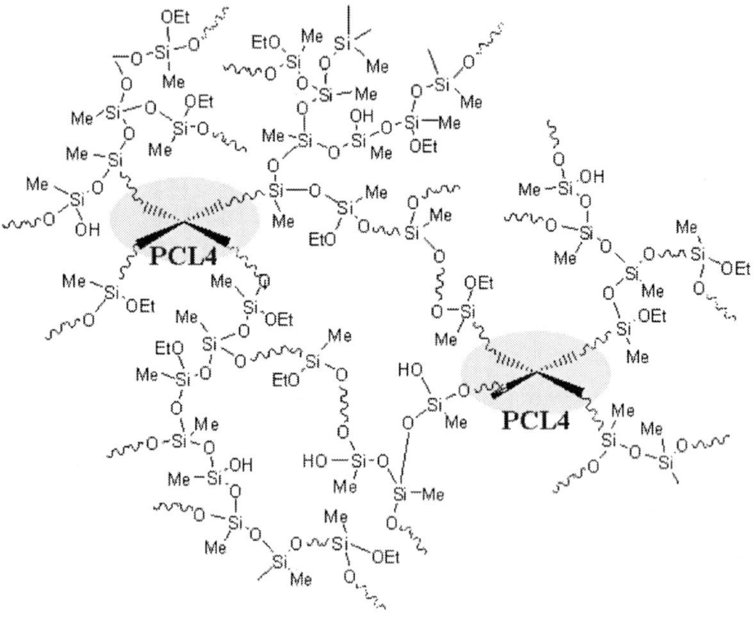

Figure 10. Polymethylsilsesquioxane-four-armed poly(ε-caprolacone) (PMSSQ-PCL4) hybrid system: PMSSQ, dielectric precursor part; PCL4, thermally labile porogen part.

In addition to the low-$k$ nanoporous PASSQ dielectrics described above, there have been attempts to prepare low-$k$ nanoporous organic polymer dielectrics based on high temperature polymers such as polyimides[113-118] and polyphenylquinoxalines.[119] These organic polymer dielectrics have high glass transition temperatures $T_g$ and thermal stabilities up to around 500°C. These high temperature polymers are modified in the syntheses to have thermally labile organic polymer blocks in their polymer backbones or as side chains (Figure 11); the labile blocks are PPO,[113-116] polystyrene,[117, 118] PCL,[119] poly(δ-valerolactone),[119] PMMA,[120-122] poly(acrylamide),[116] and poly(ethylene glycol-co-methyl ether methacrylate).[123] These syntheses have been extended to produce fluorinated polyimides.[124-126]

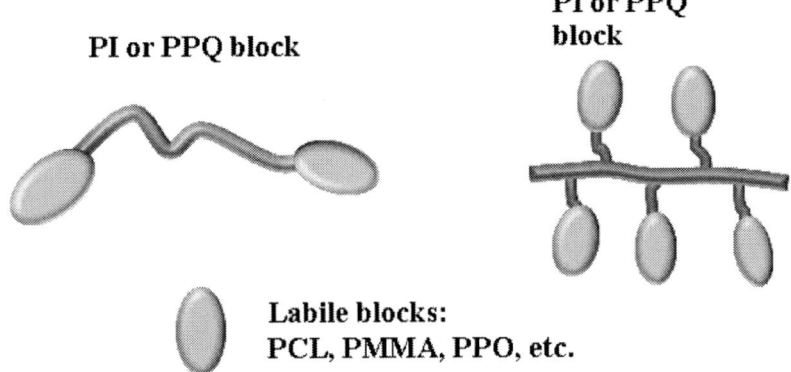

Figure 11. Organic-organic hybrid systems: triblock copolymer (top); grafting copolymer (bottom).

The labile polymer blocks undergo thermolysis below the degradation temperatures of the polyimide dielectric blocks or the poly(phenylquinoxaline) dielectric blocks, producing porous organic polymer dielectrics. The resulting porous dielectric films have porosities of 25–30 %. Low $k$ values have been achieved with these porous films.

However, the generated pores were found to collapse slowly or rapidly through thermal cycles associated with IC fabrication processes. Such pore collapses are more severe when thermal cycles are conducted above the $T_g$ of the organic polymer dielectric. Pore collapse phenomena are mainly driven by the build-up of capillary pressure inside the pore as well as by the high molecular mobility of the organic polymer dielectrics induced by thermal cycles. Pore collapse is the major drawback of this approach to producing porous organic polymer diele ctrics.

*Chapter III*

# CHARACTERIZATION OF PORE STRUCTURES

The pore structure of nanoporous organosilicates is as important for their use as low-$k$ dielectrics as their electrical, mechanical, and chemical properties[2, 6, 7, 11 3-5, 40, 42, 43, 45, 65-67, 76, 95, 101, 127, 128]. This use requires that porous low-$k$ materials have a uniform distribution of closed pores with no interconnections between pores, and with dimensions considerably smaller than the circuit feature size in order to avoid circuit defects[2, 6, 7, 11 3-5, 40, 42, 43, 45, 65-67, 76, 95, 101, 127, 128]. Accordingly, the accurate evaluation of the properties of the introduced pores is required for the successful introduction of nanoporous thin films as low-$k$ dielectrics.

As a result, much techniques have been developed for characterizing the pore structures of nanoporous low-$k$ dielectrics, such as grazing-incidence X-ray scattering (GIXS), transmission neutron/X-ray scattering (TNS/TXS) combined with specular X-ray reflectivity (SXR), transmission electron microscopy (TEM), high resolution transmission electron microscopy (HRTEM), scanning electron microscopy (SEM), field emission scanning electron microscopy (FESEM), scanning tunneling microscopy (STM), atomic force microscopy (AFM), adsorption porosimetry, ellipsometric porosimetry, positron annihilation lifetime spectroscopy (PALS).[129] They are classified into transmission radiation scattering, microscopy, porosimetry, and spectroscopy.

However, accurate measurements are difficult, because these film thickness continuously decreases (several hundred nanometers or fewer) with increasing integration and reducing feature sizes, and precise correlations between these techniques have not yet established. These problems make the selection of the appropriate method for characterization of pore structures more difficult and delay research into porogen design and pore generation methods. In this chapter, one can learn about the analytical techniques used to characterize the pore structures of nanoporous dielectric thin films and identify the strengths and weaknesses of these techniques. Especially, GIXS comes into the spotlight due to its powerful, quantitative method for the charaterization of pore structure in nanoporous low-$k$ thin film.

## III-1. GIXS

In general, TXS/TNS techniques have been used to get the information for pore structure in nanoporous materials. However, these techniques are hard to apply to pore characterization of low-$k$ materials, which are applied as form of thin film with thicknesses of several hundreds of nanometres. Recently, grazing-incidence X-ray scattering (GIXS) can be used to overcome the limitations of conventional transmission scattering techniques with respect to extremely small scattering volumes, in particular for the characterisation of the pore sizes and pore size distributions of porous thin films as well as of the thin films' properties.[3-5,74,130-132] GIXS has several important advantages over TNS and TXS: (i) a highly intense scattering pattern with high statistical significance is always obtained, even for films of nanoscale thickness; because the X-ray beam path length through the film plane is sufficiently long; (ii) there is no unfavourable scattering from the substrate on which the film is deposited; (iii) sample preparation is easy, and no additional sample treatment is required.[3, 5, 133]

The GIXS technique is very versatile and quantitative method to chracaterize the pore structure of nanoporous thin films. Besides the pore size, size distribution, and shape, the GIXS technique provides average electron density and relative porosity of film like a SXR technique.[3-5, 73, 134]

Figure 12a shows a typical GIXS geometry, here the incident X-ray beam impinges onto surface of the tin film at an angle $\alpha_i$, and the scattered pattern is recorded on a two-dimensional charge-coupled detector (2D CCD); $\alpha_f$ is the exit angle with respect to the film surface, and $2\theta_f$ is the scattering angle with respect to the plane of incidence. Porous films are generally found to have a surface roughness of a few angstroms[5], which is much smaller than a typical nanoporous film thickness, i.e. less than 800 nm, so that volumes of the interfaces with air and the silicon substrate are much smaller than that of the porous film. Thus the perturbation of the scattering due to interfacial roughness is small for porous films coated on silicon substrates.

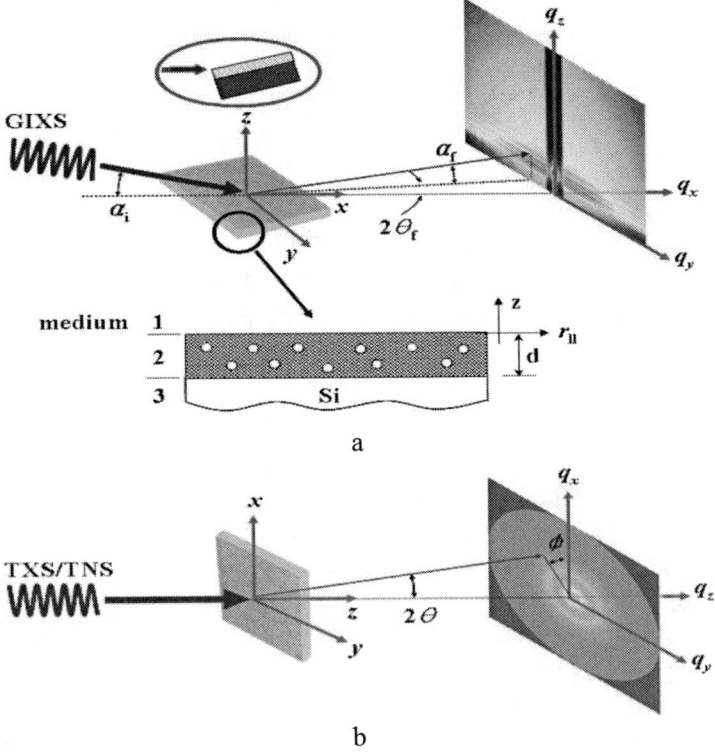

Figure 12. (a) Geometry of GIXS and schematic structural diagram of nanoporous thin film deposited onto silicon substrate: medium 1 = vacuum; medium 2 = nanoporous film; medium 3 = silicon substrate; d = thickness of medium 2 (i.e., nanoporous film). (b) Geometry of TXS. $\alpha_i$ is the incident angle at which the X-ray beam impinges on the film surface; $\alpha_f$ and $2\theta_f$ are the exit angles of the X-ray beam with respect to the film surface and to the plane of incidence, respectively, $\varphi$ is the azimuthal angle, and $q_x$, $q_y$, and $q_z$ are the components of the scattering vector q.

Contrary to transmission scattering, the analytical solution of GIXS bases on distorted wave Born approximation (DWBA) which describes the complicated reflection and refraction effects of nanoporous thin film coated onto a substrate (Figure 12a).[5, 132, 133, 135-137] Taking into consideration the negligible scattering contribution from the film surface, a novel GIXS formula with DWBA was derived to analyze quantitatively GIXS data obtained from nanoporous low-$k$ thin films[3-5, 73, 134, 138] and from thin films with other structures[133, 139-142]. The GIXS formula derivation[3-5] and its application in scattering data analyses[73, 133, 134, 138-143] have been used to show that scattering from scatterers (i.e., pores or structural elements) buried in a film coated onto a substrate results in four main types of process: (a) the incident beam scatters without reflection; (b) the scattered beam is reflected at the interface between the film and the substrate; (c) the reflected beam scatters; (d) the scattered, reflected beam is reflected once more.

A representative 2D GIXS pattern is shown in Figure 13a; it was obtained at $\alpha_i = 0.20°$ for a 108 nm thick nanoporous PMSSQ film imprinted with a 30 wt% loading of a 4-armed star-shape poly(ε-caprolactone) porogen.[73] Similar GIXS patterns were obtained for nanoporous films prepared with other porogen loadings (data not shown). PMSSQ is an excellent dielectric material because its dielectric constant (2.7–2.9) is lower than those (3.9–4.3) of silicon dioxide materials; it also has thermal stability up to 500 °C, a low moisture uptake and good mechanical strength.[3-5, 80, 144] As can be seen in Figure 13a, bright striped patterns appear along the $q_y$ direction at several exit angles ($\alpha_f$) between the critical angles of the film and the silicon substrate ($\alpha_{c,f}$ and $\alpha_{c,s}$), which arise from intense scattering due to a type of standing-wave phenomenon and total reflection at the interface between the film and the substrate. One-dimensional (1D) in-plane GIXS profiles were extracted from the measured 2D GIXS patterns along the $q_y$ direction at $\alpha_f = 0.18°$ (which is an exit angle between $\alpha_{c,f}$ and $\alpha_{c,s}$) and are plotted in Figure 14a. These scattering profiles were quantitatively analysed with the GIXS formula ($I_{\text{GIXS}}$) in the following[5]:

$$I_{GISAXS}(\alpha_f, 2\theta_f) \cong \frac{1}{16\pi^2} \cdot \frac{1-e^{-2\operatorname{Im}(q_z)\cdot d}}{2\operatorname{Im}(q_z)} \cdot \begin{bmatrix} |T_i T_f|^2 I_1(q_\|, \operatorname{Re}(q_{1,z})) + \\ |T_i R_f|^2 I_1(q_\|, \operatorname{Re}(q_{2,z})) + \\ |T_f R_i|^2 I_1(q_\|, \operatorname{Re}(q_{3,z})) + \\ |R_i R_f|^2 I_1(q_\|, \operatorname{Re}(q_{4,z})) \end{bmatrix} \quad (1)$$

where $\operatorname{Im}(q_z) = |\operatorname{Im}(k_{z,f})| + |\operatorname{Im}(k_{z,i})|$, $\operatorname{Re}(x)$ is the real part of $x$, $d$ is the film thickness, $R_i$ and $T_i$ are the reflected and transmitted amplitudes of the incoming X-ray beam respectively, $R_f$ and $T_f$ are the reflected and transmitted amplitudes of the outgoing X-ray beam respectively. In addition,

$$q_{1,z} = k_{z,f} - k_{z,i}, \quad q_{2,z} = -k_{z,f} - k_{z,i}, \quad q_{3,z} = k_{z,f} + k_{z,i},$$
$$q_{4,z} = -k_{z,f} + k_{z,i},$$

and

$$\sqrt{\phantom{xxxxxx}};$$

here, $k_{z,i}$ is the z-component of the wave vector of the incoming X-ray beam, which is given by $k_{z,i} = k_o \sqrt{n_R^2 - \cos^2 \alpha_i}$, and $k_{z,f}$ is the z-component of the wave vector of the outgoing X-ray beam, which is given by $k_{z,f} = k_o \sqrt{n_R^2 - \cos^2 \alpha_f}$, where $k_o = -2\pi/\lambda$, $\lambda$ is the wavelength of the X-ray beam, and $n_R$ is the refractive index of the film given by $n_R = 1 - \delta + i\beta$ with a dispersion $\delta$ and an absorption $\beta$, $\alpha_i$ is the out-of-plane grazing incidence angle of the incoming X-ray beam, and $\alpha_f$ is the our-of-plane exit angle of the out-going X-ray beam. $I_1$ is the scattering intensity of the porogens or pores in the film, which can be calculated kinematically.

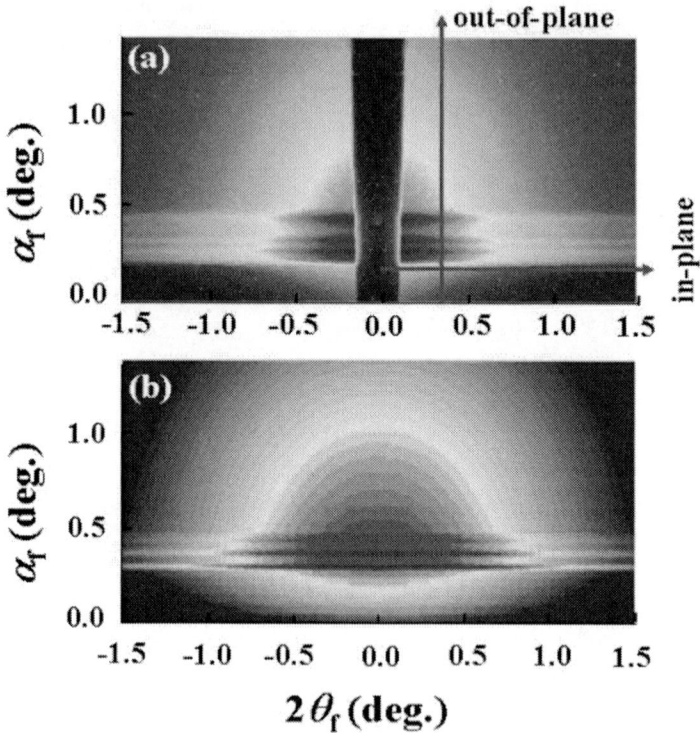

Figure 13. (a) 2D GIXS pattern measured at $\alpha_i = 0.20°$ for a 100 nm thick nanoporous low-$k$ thin film derived from a PMSSQ precursor loaded with 30 wt% 4-armed star-shaped porogen. (b) Pattern calculated for porous film in a using GIXS formula.[79] Pores were assumed to have log-normal size distribution ($r_0 = 4.46$ nm and $\sigma = 0.439$: $r_0$ and $\sigma$ are pore radius corresponding to peak maximum and width in radius distribution, respectively), and electron densities of porous film and silicon substrate are 273 and 699.5 nm$^{-3}$, respectively. The film thickness is 123 nm.

To analyze the scattering profiles using the above GISAXS formula, all scattering models (sphere, ellipsoid, cylinder and so on) for the $I_1$ term have to be examined. In this case, a sphere model is the most suitable for the structures in the films (i.e., nanopores):

$$I_1 = c \int_0^\infty n(r) v^2(r) |F(qr)|^2 S(qr) dr \qquad (2)$$

where $c$ is a constant, $v(r)$ is the volume of each pore, $F(qr)$ is the spherical form factor, and $S(qr)$ is the structure factor for the monodisperse hard sphere model[145]. $n(r)$ is the lognormal size distribution function of the pores:

$$n(r) = \frac{1}{\sqrt{2\pi} r_o \sigma e^{\sigma^2/2}} e^{\frac{-\ln(r/r_o)^2}{2\sigma^2}} \tag{3}$$

where $r$ is the pore radius, and $r_0$ and $\sigma$ are the pore radius corresponding to the peak maximum and the width of the pore radius distribution, respectively. In these fittings, all possible packing structures were further considered for the structure factor $S(qr)$ in eq. (2), and then only a randomly packed structure of spheres fits the scattering data well. Therefore the fitting results indicate that, in the porous films imprinted with 10–40wt% porogen, the spherical pores having a sharp interface with the dielectric matrix are randomly dispersed.

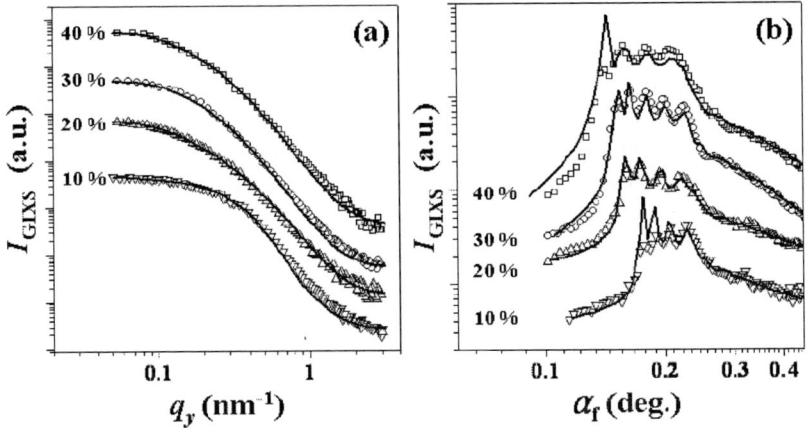

Figure 14. (a) In-plane GIXS profiles extracted along the $q_y$ direction at $\alpha_f = 0.18°$ from the 2D GIXS patterns measured for nanoporous PMSSQ films. (b) Out-of-plane GIXS profiles extracted at $2\theta_f = 0.24°$ from the 2D GIXS patterns of nanoporous PMSSQ films imprinted with the 4-armed star-shaped porogen. The symbols represent the measured data, and the solid lines were obtained by fitting the data with the GIXS formula.

This data analysis found that the film imprinted with a 10 wt% loading of porogen has pores with an average pore radius $\bar{r}$ of 3.13 nm and a relatively narrow pore size distribution ($\sigma$ = 0.395) (Figure 4). This pore size is larger than that (1.26 nm) of a single porogen molecule. As he porogen loading was increased to 40wt%, it was found that pores of larger size ($\bar{r}$ = 6.89 nm) and broader distribution ($\sigma$ = 0.450) were generated in the PMSSQ films. Collectively, these scattering data analysis results confirm that the porogen molecules have a tendency to aggregate in the PMSSQ matrix, a tendency that is enhanced as the porogen loading is increased.

Figure 14b shows the out-of-plane GIXS profiles for porous PMSSQ films prepared with porogen.[73] The out-of-plane scattering profiles of the films imprinted with 10–40wt% porogen loadings were successfully fitted with the GIXS formula and the pore parameters obtained from the analysis of the in-plane GIXS profiles, and it was confirmed that the pore size and pore size distribution are isotropic in the films.

As can be seen in Figure 14b, the profiles consist of oscillations between the critical angle of the film and that of the substrate, which are related to the film thickness. The critical angle of the film $\alpha_{c,f}$ clearly decreases with initial porogen loading, whereas that of the silicon substrate $\alpha_{c,s}$ is insensitive to porogen loading. $\alpha_{c,f}$ is directly dependent on the film electron density $\rho_e$; $\rho_e = \pi \alpha_{c,f}^2 / r_e \lambda^2$, where $r_e$ is the classical radius (2.818 × 10$^{-15}$ m) of the electron, and $\lambda$ is the wavelength of the X-ray beam used in the GIXS measurements. Thus film porosity with respect to that of a PMSSQ film prepared in the absence of porogen can be estimated from the film's electron density. The nanoporous low-$k$ films were found to have electron densities in the range 364–237 nm$^{-3}$, depending on the porogen loading. Thus, the porosity of nanoporous low-$k$ films were found to have porosities in the range 8.1–40.2 %, which were called relative porosity estimated from the electron density of the film with respect to the electron density of PMSSQ (396 nm$^{-3}$).

From the parameters determined with this procedure, 2D GIXS patterns can be generated using the GIXS formula. One of these patterns, shown in Figure 13b, was calculated for the porous film imprinted with 30 wt% porogen loading. This calculated pattern is in good agreement with the measured pattern (Figure 13a). For the porous films imprinted with 10, 20, and 30 wt% porogen loading, the calculated patterns were also found to be in good agreement with the measured patterns (data not shown).

As discussed above, the star-shape porogen was found to aggregate in the PMSSQ [4]films, ultimately imprinting nanopores that were larger than the individual molecular size. Such unfavourable aggregation of star-shape porogen molecules can be prevented partially or completely by the use of more arms in porogen molecule. One good example is the use of 6-armed star-shape poly($\varepsilon$-caprolactone) porogen.[4] More arms were found significantly to suppress aggregation in the PMSSQ throughout thin film processing.[4] Another good example is the globular ethyl acrylation of the end groups of the dendritic polypropyleneimine porogen.[3] The dendritic porogen was found to exhibit excellent miscibility with the PMSSQ precursor and was used to imprint very small nanopores with a size comparable with that of individual porogen molecules, approximately 2 nm in diameter.[3]

Thus the 2D GIXS technique is a very powerful tool for characterising the pore shape, size, size distribution, electron density, and porosity of nanoporous low-$k$ thin films of nanoscale thickness.

Moreover, GIXS measurements can be conducted in-situ as a function of thin film processing parameters (e.g., time, temperature etc.). This in situ GIXS technique can make it possible to determine the mechanism of pore generation within a thin film. One example is discussed here. For blend films of PMSSQ precursor and 4-armed star-shape porogen, in-situ GIXS measurements were performed during both heating up to 400 °C and subsequent cooling to room temperature.[73] Of the GIXS data obtained as a function of temperature and time, some representative 2D scattering patterns and 1D in-plane and out-of-plane scattering profiles are presented in Figure 15. As can be seen in Figure 15a, b, and c, all the 1D scattering profiles were well fitted with the GIXS formula for hard spherical particles with a lognormal size distribution. The pore sizes of the porogen and the imprinted nanopores are plotted in Figure 16a as a function of the temperature during the heating run as well as of the initial porogen loading. Among them, for the PMSSQ dielectric film loaded with 30 wt% 4-armed star-shaped porogen, pore radii and their radius distributions determined from the GIXS analysis of the in-plane scattering profile data for in Figures 15a and b are plotted in Figure 16b.

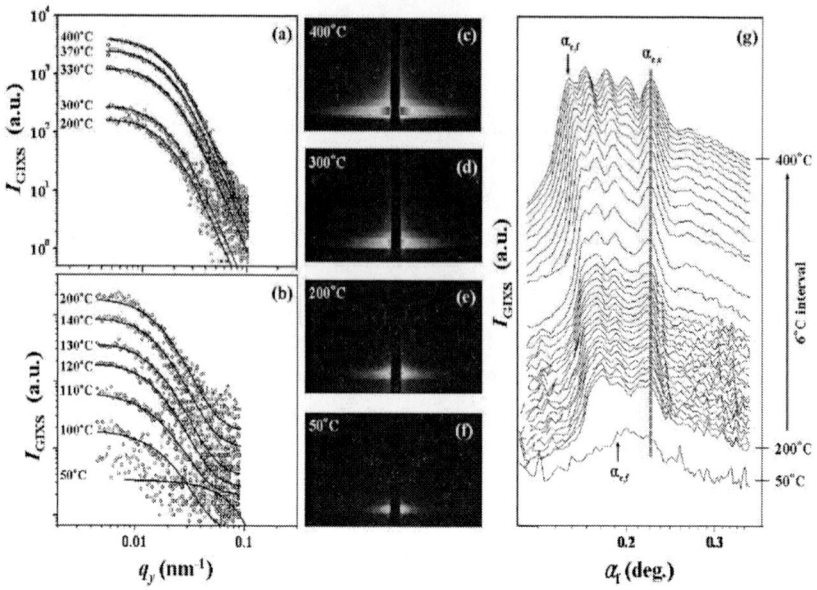

Figure 15. (a), (b) In-plane GIXS profiles at $\alpha_f = 0.18°$ of 2D GIXS patterns measured during heating (2.0 °C min$^{-1}$) of a PMSSQ precursor film loaded with 30 wt% 4-armed star-shaped porogen under vacuum. 2D GIXS pattern measured during heating (2.0 °C min$^{-1}$) of a PMSSQ precursor film: (c) 400 °C; (d) 300 °C; (e) 200 °C; (f) 50 °C. (g) Out-of-plane GIXS profiles at $2\theta_f = 0.24°$ of 2D GIXS patterns measured during heating (2.0 °C min$^{-1}$) of a PMSSQ precursor film loaded with 30 wt% 4-armed star-shaped porogen under vacuum. $\alpha_f$ indicates the critical angle of the composite film.

This in situ GIXS study obtained the following results. Heating of the porogen-loaded PMSSQ precursor matrix produces a curing reaction in the precursor matrix component, resulting in the phase separation of the porogen component. On heating, limited aggregations of the porogen, however, took place in only a small temperature range of 100–140 °C as a result of phase separation induced by the competition of the curing and hybridization reactions of the dielectric precursor and porogen (Figure 16); higher porogen loading resulted in relatively large porogen aggregates and a greater size distribution. The developed porogen aggregates underwent thermal firing above 300 °C without further growth and movement, and ultimately left their individual footprints in the film as spherical nanopores.

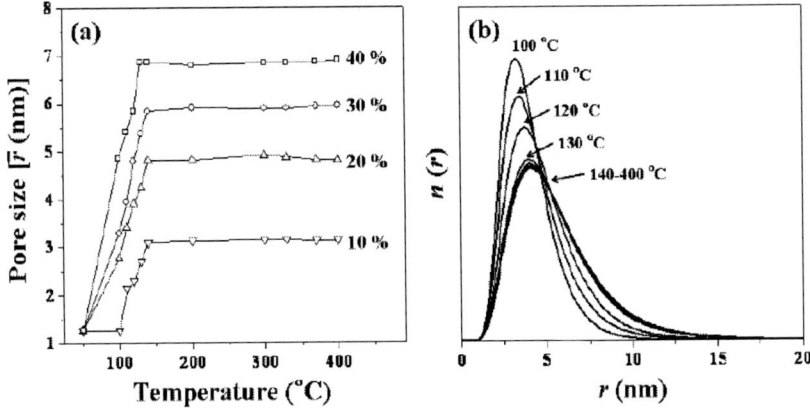

Figure 16. (a) Average radii of porogens and pores in the PMSSQ films determined from the GIXS analysis of the in-plane scattering profile data obtained for the films loaded with 10-40 wt% 4-armed star-shaped porogen. (b) Porogen and pore radii and their radius distributions determined from the GIXS analysis of the in-plane scattering profile data for the PMSSQ dielectric film loaded with 30 wt% 4-armed star-shaped porogen in Figures 15a and b.

Figure 15g shows representative out-of-plane scattering profiles, which were extracted from the 2D GISAXS patterns measured during the heating of a PMSSQ precursor composite film with a 30 wt% porogen loading. Here, a variation in the critical angle of the composite film, $\alpha_{c,f}$, with increasing temperature is clearly evident. The $\alpha_{c,f}$ value of 0.186° ($\rho_e = 401$ nm$^{-3}$) at 50 °C shows a very slow shift towards the low-angle region as the temperature draws closer to 300 °C. This is due to the removal of the water and ethyl alcohol byproducts, formed in the composite film during the curing of the precursor matrix component. Above 300 °C, the $\alpha_{c,f}$ shows a dramatic shift towards the low-angle region as the temperature is increased, achieving a final value of 0.152° ($\rho_e = 237$ nm$^{-3}$) at 400 °C. This drastic shift in $\alpha_{c,f}$ towards the low-angle region is attributed primarily to the electron density of the film, which is lowered when pores are generated in the film through the thermal degradation of the porogen aggregates, and partly lowered following removal of the water and ethyl alcohol byproducts from the hybrid film. Similar trends in the $\alpha_{c,f}$ variation with temperature were observed in the out-of-plane scattering profiles of the other hybrid films (data not shown).

## III-2. Transmission Radiation Scattering

Transmission neutron and X-ray scattering (TNS and TXS) are widely used in conjunction with specular X-ray reflectivity (SXR) to obtain the average pore sizes and size distributions of porous materials (Figure 12b).[128, 146-150] In TNS, neutron beams are scattered when they travel through media with varying neutron scattering length densities (SLDs). The key parameter determining the scattering intensity is contrast, which is the difference between the SLDs of the pores and the matrix. The scattering intensity is mostly interpreted by the correlation length approach with the random two-phase model of Debye.[148, 151, 152] However, it is difficult to obtain pore structure information with such methods when there is a lack of contrast between the pores and the matrix, which is always attributed to the very thin film thicknesses of ≤ 800 nm, because a significant contrast term is crucial to obtaining sufficient scattering intensity. One method for overcoming the lack of contrast to some extent in such measurements is to employ a stack of several tens of thin films deposited on thin silicon wafers, which have low attenuation.[148, 150]

A TXS investigation of porous polymethylsilsesquioxane (PMSSQ) thin films has been reported in which the TXS data were analysed with a hard sphere model.[5, 134, 153] In this case, the scattering contrast between the pores and the MSSQ matrix was sufficient. This approach has also been used to characterise the three-dimensional disordered morphologies of isotropic two-phase materials, and to study the transition from closed pores to interconnected pores or to a bicontinuous morphology that arises with increases in the content of the pore generator.[154] TXS provides information about the mean pore size and size distribution and can be carried out with various models (e.g., sphere, core shell, disc and rod etc.).

Along with transmission scattering technique, SXR can give detailed structural information about the low-$k$ dielectric thin films in terms of porosity, electron density/or mass density, density distribution, thickness, out-of-plane coefficient of thermal expansion (CTE), roughness of surface and interface.[65, 138, 146-148, 155-157] Using SXR, the electron density ($\rho_e$) of a thin film can be determined independently from the film thickness. Figure 17 diaplays a representative SXR profile, which was measured for a nanoporous PMSSQ low-$k$ film. The two critical angles, which are attributed to the film ($\alpha_{c,f}$) and the substrate ($\alpha_{c,s}$), respectively, can be clearly observed in the measured data. The critical angle is directly related to the average electron density of the corresponding fillm by $\alpha_{c,f} = \lambda \left( \rho_e r_e / \pi \right)^{1/2}$, where $\lambda$ represents the wavelength of radiation and $r_e$ is the classical electron radius.[135, 136] Thus the average electron densities of the film and the substrate are directly determined by the two critical angles, and they can be transformed into mass densities if the chemical compositions of the materials are known. The electron density is directly proportional to the mass density. In general, the porosity, so-called relative porosity, can be obtained by comparing the average film density to an assumed density of the non-porous film.

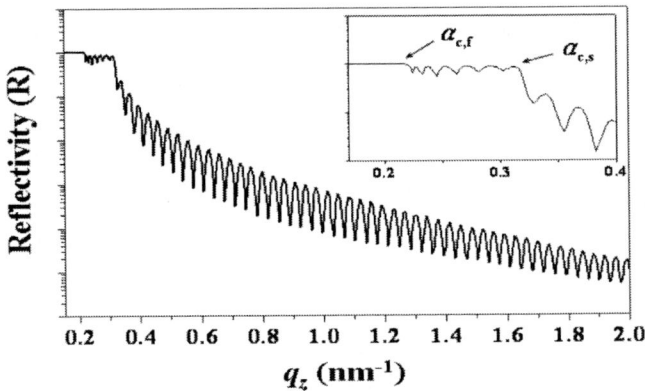

Figure 17. (a) A representative specular X-ray reflectivity curve measured from a thin nanoporous PMSSQ low-$k$ film. $\alpha_{c,f}$ and $\alpha_{c,s}$ are the critical angles of the film and the silicon substrate, respectively. The inset shows a magnification of the region around the two critical angles.

However, for low scattering volumes, such as those of thin films, TXS is no longer applicable owing to its low sensitivity and resolution. Further, it requires an intense, high-energy (approximately 15 keV) X-ray beam to produce an acceptable scattering signal, as the X-rays must pass through both the film and the much thicker substrate on which the film is deposited.[5] In addition, for TXS to be effective, the substrate should be free of crystal domain boundaries, as these can give rise to unfavourable X-ray scattering and reflection.[5]

## III-3. MICROSCOPY

Several microscopy techniques, such as such as TEM, HRTEM, SEM, FESEM, STM, and AFM are are used to characterise pore structures.[44, 128, 153, 158-161] The spherical structure and the randomly dispersed pores in porous PMSSQ dielectric film is shown in Figure 18. This TEM image clearly shows that the pores generated in the dielectric film by the sacrificial thermal degradation of the porogen are spherical in shape. [4, 134] Recently, High-resolution electron microscopy has recently been used to produce images with atomic resolution.[162] These techniques are widely used for the direct characterisation of pore structures because of their powerful visualisation features, which can provide topographical and structural information in plan or cross-sectional views. Further, qualitative analysis is possible with these techniques, even with specimens of limited area. However, they are insensitive and complicated to use in quantitative analysis, because of the overlapping of pores in crosssectional views, and have profound analytic resolution limitations. Moreover, TEM and SEM techniques require intensive and exacting specimen preparation processes, typically under high vacuum conditions and with highly focused ion beams. Nevertheless, they will continue to be widely used in conjunction with other techniques.

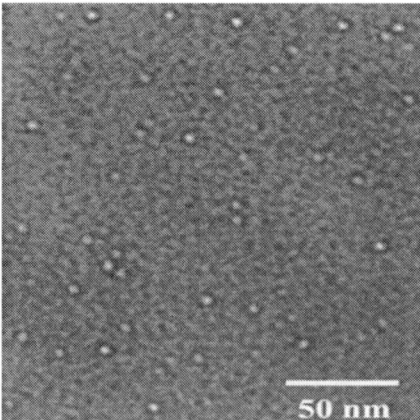

Figure 18. TEM image of a nanoporous PMSSQ dielectric prepared from a PMSSQ precursor sample loaded with 10 wt% 4-armed star-shaped porogen.

## III-4. POROSIMETRY

Adsorption porosimetry has been widely used to determine pore size distribution and porosity; this method determines the adsorption–desorption isotherm of nitrogen vapor for the sample by weighing the sample.[163, 164] However, there are some problems in the adaptation of this technique to the requirements of industry. First, most current research is focused on materialising the nanoporous materials as thin films with thicknesses of several hundreds of nanometres. However, thin films have insufficient mass for adequate measurements with the traditional nitrogen adsorption technique, because it is based on the use of a microbalance, which is only suitable for samples with relatively high mass. Secondly, it is only applicable to open pores, not to closed pores. Thirdly, it is destructive, because the sample must be grind into a powder to be weighed with the microbalance. However, for accurate results to be obtained, the thin film sample must retain the same structural features that it had when it was produced. Finally, adsorption porosimetric measurements are usually performed near the boiling point of nitrogen vapour (around −195.6 °C). These unfavourable conditions act as crack-driving forces. In an attempt to investigate thin film specimens non-destructively at room temperature, and thus to avoid severe stress conditions, the use of quartz crystal microbalance (QCM)[163, 165] and surface acoustic wave (SAW) methods[166] has been suggested. However, these methods cannot be used for sensing the adsorption of vapor onto a silicon wafer or other substrate, because QCM and SAW require that the sample be deposited directly onto the sensor.

To overcome the significant drawbacks of gas adsorption porosimetry, ellipsometric porosimetry, which opens up a new genre of adsorption porosimetry, has been introduced.[167, 168] This technique is similar to gas adsorption porosimetry in that the sample is exposed to an adsorptive gas or probe liquid (e.g., toluene, heptane etc.). The change in the refractive index of the sample is then employed for the determination of the mass of the adsorptive that is condensed and/or adsorbed into pores, instead of the direct weighing used in adsorption porosimetry. The full porosity, open pore volume and pore interconnectivity can be calculated from the changes in the refractive index and the thickness of the film.[169, 170] It is also possible, with this method, to characterise a film's surface roughness (which includes open pores at the surface)[171]

The reliability of pore size distribution analysis with ellipsometric porosimetry using various organic adsorptives has been demonstrated by Baklanov *et al.*[167] Although the use of ellipsometric porosimetry can provide impressive physical and structural details, this technique can only be used for open pores, as is the case for nitrogen porosimetry. Further, this approach assumes that the refractive index of liquid condensed and/or adsorbed in nanoscale pores is identical to that of the bulk liquid. This assumption is only reasonable when the pore size approaches that of a few adsorptive molecules. Lastly, this technique cannot take into account the swelling of the thin film by the probe liquid, which results in an overestimate of the gas uptake and thus of the pore size.

## III-5. SPECTROSCOPY

Positron annihilation lifetime spectroscopy (PALS) is applicable to the determination of free volume at the atomic or molecular level and is based on the annihilation lifetime of positronium (Ps, positron-electron bound state) through collisions with the electrons of the pores' inner surfaces (Figure 19).[172-174] Unlike adsorption techniques, this method can be used for both closed pores and open pores. Gidley *et al.* have sucessfully demonstrated that pore structure (pore size, size distribtuion, and interconnectivity) of both isolated pores and highly interconnected open pores, in the latter case, thin capping layer such as Al on film surface is necessary to Keep Ps escaping into vacuum (on the right side of Figure 19).[175-178] Moreover, the PALS can be applied a multilevel thin film which the film is capped with a diffusion barrier (e.g., Ta, TaN or TiN) or sacrificial layer such as silicon oxide layer prepared from tetraethoxysilane (TEOS).[168] These features resolve the major problems of porosimetry methods. Further, the possibility of analysis of capped systems enables the use of this method on practical multilevel integrated circuits with various layers such as an etch stop layer, a capping layer and a copper conductor with a barrier/metal seed layer.[178]

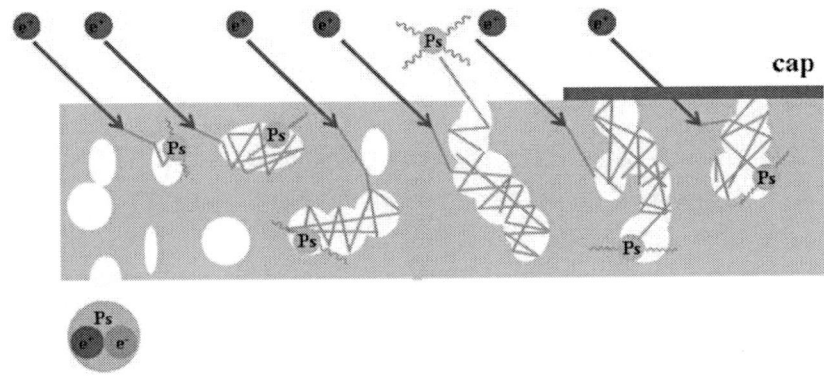

Figure 19. Annihilation of positron in two types of nanoporous low-$k$ thin film: (left) closed pores; (right) interconnected open pores. $e^+$s are positrons and they form Ps (positron –electron bound state) within pores.

The most powerful feature of PALS is its sensitivity to much smaller pores (down to a few angstroms) than those detectable with TEM, TNS and TXS (down to a few nanometres). Recently, the complete depth-dependent pore structure of a nanoporous thin film was characterised using depth-profiled PALS.[175-178] This technique is suitable to monitor any variations in the depth-dependent pore structure in the fabrication of a nanoporous low-$k$ thin film.[169, 171] To calculate the pore size distribution from the lifetime spectra, a calibration curve is required that connects the Ps lifetime with the pore size.[128]

The PALS method does have some limitations.[176, 179-181] First, if the pores in the film system under investigation are open and highly interconnected, the highly mobile Ps diffuses out of the film and escapes into the vacuum, resulting in a lifetime the same as that in a vacuum (about 142 ns), and then information about pore size cannot be obtained. This problem with open, interconnected pores can be solved by a capping layer being deposited onto the thin film layer to prevent the Ps escaping from the film layer into the vacuum. Secondly, in some materials (e.g., polyimides), Ps formation is suppressed because of an abundance of free radicals that scavenge electrons during the Ps formation process. Thirdly, the Ps formed within the thin film must be able to diffuse into the pores; otherwise PALS probes only molecular voids (i.e., molecular-free volumes) in the bulk film part rather than the pores of interest in the nanoporous film. Finally, commercial positron beams are not readily available.

## III-6. COMPARITIVE STUDIES OF CHARACTERIZATION OF PORE STRUCTURE

As mentioned above, various methods have been utilized for characterising pore structure of nanoporous low-$k$ thin films: microscopy (TEM, HRTEM, SEM, FESEM, STM, and AFM), porosimetry (adsorption porosimetry and ellipsometric porosimetry), spectroscopy (PALS), and radiation scattering (TXS, TNS, SXR, and GIXS). It is difficult to select which technique is the best tool for the characterisation of nanoporous low-$k$ thin films, because all methods have strength and weakness according to their lights based on physicochemical principles and methodologies. Furthermore, all the techniques except for the microscopy methods, the pore size and pore size distribution of nanoporous thin films are not directly obtained from raw data. Information about the pore size and size distribution of a thin film can only be obtained from the raw data by formulation of a model that takes into account the pore structure and the pore–probe (e.g., laser, positron, X-ray, and neutron) interaction. This strong dependence on the assumed model may lead to differences between the results of the various techniques. Thus, one need to be careful in the selection of technique but also to optimise the selected technique for the specific task. Further more, comparative studies are required to establish criteria for characterization of nanoporous low-$k$ thin film. Several comparative studies of various techniques have been reported.[131, 146, 148, 152, 161, 169, 176, 182-184] These comparative studies found that the pore sizes obtained with the various techniques were in reasonable agreement. However, detailed correlations between the pore sizes and size distributions obtained with the various methods were not given.

*Chapter IV*

# CONCLUSIONS

As reviewed above, there are two principal methods for producing low-$k$ spin-on dielectric materials containing closed nanopores, which are the most promising interdielectric layers for the production of advanced ICs by the semiconductor industry.

The first approach is to prepare thermally and dimensionally stable hollow closed nanoparticles with a low $k$ value and disperse them uniformly throughout the dielectric film volume, producing nanoporous low-$k$ thin films. In this case, avoiding or minimizing the aggregation of hollow nanoparticles within the dielectric film is preferred but not absolutely required because the hollow particles are thermally and dimensionally stable, and thus stand alone individually. Moreover, due to their high thermal and dimensional stability, there are a wide variety of dielectric matrices that can be used with these hollow nanoparticles, including curable and noncurable dielectrics and precursors as well as inorganic and organic dielectrics. However, the fabrication of hollow nanoparticles of the desired size and miscibility with dielectric matrices remains a significant challenge.

The second approach is based on the addition to the dielectric precursor of nanoscale porogens with tailored thermal stabilities. The stability of these porogens must be such that they are not affected by the coating and drying steps in the dielectric film formation process; they are removed by sacrificial thermal degradation during the final heat treatment of the dielectric films at temperatures typically in the range 300–400 °C. Their volume distribution in the film is the template for the formation of residual closed nanopores in the dielectric film. The porous fraction of the film is in principle directly related to the fraction of porogen with respect to the total solids in the dielectric precursor solution, and the size of the sacrificial porogens is directly related to the final pore size. However, there are some practical requirements if these relationships between the sacrificial porogens, imprinted pores and porosity are to be maintained. Firstly, the sacrificial porogen should be compatible with the dielectric matrix material in order to avoid porogen aggregation. Secondly, the sacrificial porogens should be uniformly distributed throughout the film volume in order to avoid the coalescence or interconnection of the pores. There are two ways in which the sacrificial porogens can be introduced into the dielectric precursor solution. One method is the dispersion of porogens in the solution. The second is chemically linking the sacrificial porogens to the network polymers as block components of the backbone or through grafting. This second method enables control of the volume distribution of the porogens in the dielectric film. Finally, the porogen template approach requires that the dielectric matrix film has a higher degree of cross-linking so that it is dimensionally stable when the pores are created. The porous structure is therefore less affected by any further processes associated with the resulting nanoporous dielectric film; this is why silicate and organosilicate precursors (e.g., PASSQ precursors) and their cured dielectrics are appropriate for the porogen template approach to producing low-$k$ nanoporous dielectrics. Otherwise the pores in the dielectric film may collapse during post-processing of the film, including thermal cycles, which arises because of high capillary pressure and the molecular mobility of the film induced by thermal processing; this is why non-curable organic polymers (e.g., polyimides and polyphenylquinoxalines) are not appropriate for the porogen template approach. In addition, closed nanoporosity can be obtained, within certain boundary conditions related to the nature and total load of the porogen.

The closed pores in a porous low-$k$ dielectric layer film must be 5–10 times smaller than the smallest device feature; the minimum metal feature size is nowadays approaching 50 nm and may reach 25 nm in the near future.

Nanoporous materials for low-$k$ application need control of pore size and size uniformity to nanometer scale. Because the mechanical stability of nanoporous materials for integration to multilavel semiconductor are close related to the pore structures. Therefore, the pore structures of nanoporous materials are very important for their use as low-$k$ dielectrics as their electrical, mechanical, and chemical properties, and the accurate evaluation of the properties of the introduced pores is required for the successful introduction of nanoporous thin films as low-$k$ dielectrics. However, the pore size and film thickness of interlayer dielectrics (ILDs) continue to be reduced in the pursuit of increased integration and reduced feature sizes. Thus the characterization of pore structure becomes more and more difficult. For this reason, various advanced techniques for the characterisation of pore structures have recently been developed: GIXS, TNS/TXS combined SXR, various kinds of microscopy, adsorption porosimetry, ellipsometric porosimetry, and PALS. In this book, usefulness of these techniques in the precise, and quantitative characterization of the pore shapes, size and size distribution of nanoporous low-$k$ thin film were mentioned. Characterization of structures with dimensions approaching those of atoms is required. However, no all-encompassing technique is currently available. Thus a synergistic approach, i.e., the application in combination of various analytical techniques to obtain results superior to those obtained with each individual technique, is necessary.

In conclusion, considering the Semiconductor Industry Association's International Technology Roadmap for Semiconductors, significant challenges remain in the fabrication and production of high performance low-$k$ dielectric materials consisting of closed nanopores of 4 nm or less that meet the requirements of the production of advanced ICs in the microelectronics industry. Moreover, suitable characterization technique for precise and quantitative evaluation of pore structures is necessary to advance the development of nanoporous low-$k$ materials with porogen and pore generation method points of view.

This work was supported supported by the National Research Foundation (NRF) of Korea (National Research Lab Program and Center for Electro-Photo Behaviors in Advanced Molecular Systems) and the Ministry of Education, Science & Technology (MEST) (BK21 Program and World Class University Program).

# REFERENCES

[1] Ree, M.; Goh, W. H.; Kim, Y. *Polymer Bulletin* 1995, 35 (1-2), 215-222.
[2] Miller, R. D. *Science* 1999, 286 (5439), 421-423.
[3] Lee, B.; Park, Y. H.; Hwang, Y. T.; Weontae, O. H.; Yoon, J.; Ree, M. *Nat. Mater.* 2005, 4, (2), 147-151.
[4] Lee, B.; Oh, W.; Hwang, Y.; Park, Y. H.; Yoon, J.; Jin, K. S.; Heo, K.; Kim, J.; Kim, K. W.; Ree, M. *Adv. Mater.* 2005, 17 (6), 696-701.
[5] Lee, B.; Yoon, J.; Oh, W.; Hwang, Y.; Heo, K.; Jin, K. S.; Kim, J.; Kim, K. -W.; Ree, M. *Macromolecules* 2005, 38 (8), 3395-3405.
[6] Morgen, M.; Ryan, E. T.; Zhao, J. H.; Hu, C.; Cho, T.; Ho, P. S. *Annu. Rev. Mater. Sci.* 2000, 30, 645-680.
[7] Maex, K.; Baklanov, M. R.; Shamiryan, D.; Iacopi, F.; Brongersma, S. H.; Yanovitskaya, Z. S. *J. Appl. Phys.* 2003, 93 (11), 8793-8841.
[8] Ree, M.; Yoon, J.; Heo, K. *J. Mater. Chem* 2006, 16, 685-697.
[9] Czornyj, G.; Chen, K. J.; Prada-Silva, G.; Arnold, A.; Souleotis, H. A.; Kim, S.; Ree, M.; Volksen, W.; Dawson, D.; DiPietro, R. *Proc. Elect. Comp. Tech. (IEEE)* 1992, 42, 682-692.
[10] International Technology Roadmap for Semiconductors 1999.
[11] Maier, G. *Prog. Polym. Sci.* 2001, 26 (1), 3-65.
[12] Ree, M.; Shin, T. J.; Lee, S. W. *Korea Polymer J.* 2001, 9 (1), 1-19.
[13] Kim, Y.; Goh, W. H.; Chang, T.; Ha, C. S.; Ree, M. *Adv. Eng. Mater.* 2004, 6 (1), 39-43.
[14] Shin, T. J.; Ree, M. *Langmuir* 2005, 21 (13), 6081-6085.

[15] Yu, J.; Ree, M.; Shin, T. J.; Park, Y. H.; Cai, W.; Zhou, D.; Lee, K. W. *Macromol. Chem. Phys.* 2000, 201 (5), 491-499.
[16] Kim, S. I.; Shin, T. J.; Ree, M.; Lee, H.; Chang, T.; Lee, C.; Woo, T. H.; Rhee, S. B. *Polymer* 2000, 41 (14), 5173-5184.
[17] Yu, J.; Ree, M.; Shin, T. J.; Wang, X.; Cai, W.; Zhou, D.; Lee, K. W. *J. Polym. Sci.: Part B: Polym. Phys.* 1999, 37 (19), 2806-2814.
[18] Pyo, S. M.; Kim, S. I.; Shin, T. J.; Park, Y. H.; Ree, M. *J. Polym. Sci.: Part A.: Polym. Chem.* 1999, 37 (2-7), 937-957.
[19] Kim, Y.; Kang, E.; Kwon, Y. S.; Cho, W. J.; Cho, C.; Chang, M.; Ree, M.; Chang, T.; Ha, C. S. *Synthetic Metals* 1997, 85 (1-3), 1399-1400.
[20] Kim, Y.; Lee, W. K.; Cho, W. J.; Ha, C. S.; Ree, M.; Chang, T. *Polym. Int.* 1997, 43 (2), 129-136.
[21] Ree, M.; Kim, K.; Woo, S. H.; Chang, H. *J. Appl. Phys.* 1997, 81 (2), 698-708.
[22] Kim, Y.; Ree, M.; Chang, T.; Ha, C. S.; Nunes, T. L.; Lin, J. S. *J. Polym. Sci.* 1995, 33, 2075.
[23] Ree, M.; Goh, W. H.; Park, J. W.; Lee, M. H.; Rhee, S. B. *Polymer Bulletin* 1995, 35 (1-2), 129-136.
[24] Kim, Y.; Ree, M.; Chang, T.; Ha, C. S. *Polymer Bulletin* 1995, 34 (2), 175-182.
[25] Ree, M.; Han, H.; Gryte, C. C. *J. Polym. Sci., Polym. Phys. Ed.* 1995, 33, 505-516.
[26] Ree, M.; Nunes, T. L.; Chen, K.-J. R. *J. Polym. Sci.: Part B: Polym. Phys.* 1995, 33, 453-465.
[27] Ree, M.; Han, H.; Gryte, C. C. *High Perform. Polym.* 1994, 6, 325-325.
[28] Ree, M.; Nunes, T. L.; Lin, J. S. *Polymer* 1994, 35 (6), 1148-1156.
[29] Ree, M.; Chu, C. W.; Goldberg, M. J. *J. Appl. Phys.* 1994, 75 (3), 1410-1419.
[30] Ree, M.; Swanson, S.; Volksen, W. *Polymer* 1993, 34 (7), 1423-1430.
[31] Ree, M.; Chen, K. K.-J. R.; Czornyj, G. *Polym. Eng. Sci.* 1992, 32, 924-924.
[32] Ree, M.; Chen, K. K.-J. R.; Kirby, D. P.; Katzenellenbogen, N.; Grischkowsky, D. *J. Appl. Phys.* 1992, 72 (5), 2014-2021.
[33] Rojstaczer, S.; Ree, M.; Yoon, D. Y.; Volksen, W. *J. Polym. Sci.: Part B: Polym. Phys.* 1992, 30 (2), 133-143.

[34] Robertson, W. M.; Arjavalingam, G.; Hougham, G.; Kopesav, G. V.; Edelstein, D.; Ree, M. H.; Chapple-Sokol, J. D. *Electron. Lett.* 1992, 28 (1), 62-63.

[35] Ree, M.; Nunes, T. L.; Czornyj, G.; Volksen, W. *Polymer* 1992, 33 (6), 1228-1236.

[36] Ree, M.; Yoon, D. Y.; Volksen, W. *J. Polym. Sci.: Part B: Polym. Phys.* 1991, 29 (10), 1203-1213.

[37] Moylan, C. R.; Best, M. E.; Ree, M. *J. Polym. Sci.: Part B: Polym. Phys.* 1991, 29 (1), 87-92.

[38] Ree, M.; Shin, T. J.; Park, Y. H.; Lee, H.; Chang, T. *Korea Polymer J.* 1999, 7 (6), 370-376.

[39] Goh, W. H.; Kim, K.; Kim, S. I.; Shin, T. J.; Ree, M. *Korea Polymer J.* 1998, 6 (3), 241-248.

[40] Azooz, M. A.; Hwang, Y. T.; Ree, M. *Egypt. J. Chem.* 2003, 46, 741-756.

[41] Shin, T. J.; Ree, M. *Macromol. Chem. Phys.* 2002, 203 (5-6), 791-800.

[42] Oh, W.; Ree, M. *Langmuir* 2004, 20 (16), 6932-6939.

[43] Nguyen, C. V.; Carter, K. R.; Hawker, C. J.; Hedrick, J. L.; Jaffe, R. L.; Miller, R. D.; Remenar, J. F.; Rhee, H. W.; Rice, P. M.; Toney, M. F.; Trollsas, M.; Yoon, D. Y. *Chem. Mater.* 1999, 11 (11), 3080-3085.

[44] Hedrick, J. L.; Miller, R. D.; Hawker, C. J.; Carter, K. R.; Volksen, W.; Yoon, D. Y.; Trollsas, M. *Adv. Mater.* 1998, 10 (13), 1049-1053.

[45] Huang, E.; Toney, M. F.; Volksen, W.; Mecerreyes, D.; Brock, P.; Kim, H. C.; Hawker, C. J.; Hedrick, J. L.; Lee, V. Y.; Magbitang, T.; Miller, R. D.; Lurio, L. B. *Appl.Phys. Lett.* 2002, 81 (12), 2232.

[46] Mukherjee, S. P.; Suryanarayana, D.; Strope, D. H. *J. Non-Crystalline Solids* 1992, 147-48, 783-791.

[47] Muktherjee, S. P.; Cordars, J. F.; Debsikdar, J. C. *Adv. Ceram. Mater.* 1988, 3 (5), 463-480.

[48] Chandrashekhar, G. V.; Shafer, M. W. *Mater. Res. Soc. Symp. Proc.* 1986, 72, 309-315.

[49] Fischer, M.; Vogtle, F. *Angew. Chem. Int. Ed.* 1999, 38 (7), 884-905.

[50] Narayanan, V. V.; Newkome, G. R. *Top. Curr. Chem.* 1998, 197, 19-77.

[51] Zimmerman, S. C.; Wendland, M. S.; Rakow, N. A.; Zharov, I.; Suslick, K. S. *Nature* 2002, 418 (6896), 399-403.
[52] Tomalia, D. A.; Baker, H.; Dewald, J.; Hall, M.; Kallos, G.; Martin, S.; Roeck, J.; Ryder, J.; Smith, P. *Polymer J.* 1984, 17 (1), 117-132.
[53] De Brabander-van Den Berg, E. M. M.; Meijer, E. W. *Angew. Chem. Int. Ed.* 1993, 32 (9), 1308-1311.
[54] Jikei, M.; Kakimoto, M. A. *J. Polym. Sci., Part A: Polym. Chem.* 2004, 42 (6), 1293-1309.
[55] Kim, Y. H. *J, Polym. Sci.:Part A: Polym.Chem.* 1998, 36 (11), 1685-1698.
[56] Oh, W.; Shin, T. J.; Ree, M.; Jin, M. Y.; Char, K. *Mol. Cryst. Liq. Crys. Sci. Tech. Sec. A: Mol. Cryst. Liq.Cryst.* 2001, 371, 397-402.
[57] Baney, R. H.; Itoh, M.; Sakakibara, A.; Suzuki, T. *Chem. Rev.* 1995, 95 (5), 1409-1430.
[58] Shin, Y. C.; Choi, K. Y.; Jin, M. Y.; Hong, S. K.; Cho, D.; Chang, T.; Ree, M. *Korea Polymer J.* 2001, 9 (2), 100-106.
[59] Hirao, A.; Hayashi, M.; Loykulnant, S.; Sugiyama, K.; Ryu, S. W.; Haraguchi, N.; Matsuo, A.; Higashihara, T. *Prog. Polym. Sci.* 2005, 30 (2), 111-182.
[60] Daoud, M.; Cotton, J. P. *J. phy. Paris* 1982, 43 (3), 531-538.
[61] Ishizu, K.; Ono, T.; Uchida, S. *Macromol. Chem. Phys.* 1997, 198 (10), 3255-3265.
[62] Hsu, H. P.; Nadler, W.; Grassberger, P. *Macromolecules* 2004, 37 (12), 4658-4663.
[63] Zimm, B. H.; Stockmayer, W. H. *J. Chem. Phys.* 1949, 17 (12), 1301-1314.
[64] Witten, T. A.; Pincus, P. A.; Cates, M. E. *Europhys. Lett.* 1986, 2 (2), 137-140.
[65] Bolze, J.; Ree, M.; Youn, H. S.; Chu, S. H.; Char, K. *Langmuir* 2001, 17 (21), 6683-6691.
[66] Oh, W.; Hwang, Y.; Park, Y. H.; Ree, M.; Chu, S. H.; Char, K.; Lee, J. K.; Kim, S. Y. *Polymer* 2003, 44 (8), 2519-2527.
[67] Hedrick, J. L.; Miller, R. D.; Hawker, C. J.; Carter, K. R.; Volksen, W.; Yoon, D. Y.; Trollsas, M. *Adv. Mater.* 1998, 10 (13), 1049-1053.
[68] Trollsas, M.; Hedrick, J. L.; Mecerreyes, D.; Dubois, P.; Jerome, R.; Ihre, H.; Hult, A. *Macromolecules* 1997, 30 (26), 8508-8511.

[69] Trollsas, M.; Hedrick, J. L.; Mecerreyes, D.; Dubois, P.; Jerome, R.; Ihre, H.; Hult, A. *Macromolecules* 1998, 31 (9), 2756-2763.

[70] Trollsas, M.; Hedrick, J. L. *J. Am. Chem. Soc.* 1998, 120 (19), 4644-4651.

[71] Heise, A.; Nguyen, C.; Malek, R.; Hedrick, J. L.; Frank, C. W.; Miller, R. D. *Macromolecules* 2000, 33 (7), 2346-2354.

[72] Jin, K. S.; Heo, K.; Oh, W.; Yoon, J.; Lee, B.; Hwang, Y.; Kim, J. S.; Park, Y. H.; Chang, T.; Ree, M. *J. Appl. Crystallog.* 2007, 40, s631-s636.

[73] Yoon, J.; Heo, K.; Oh, W.; Jin, K. S.; Jin, S.; Kim, J.; Kim, K.-W.; Chang, T.; Ree, M. *Nanotechnology* 2006, 17, 3490-3498.

[74] Lee, B.; Oh, W.; Yoon, J.; Hwang, Y.; Kim, J.; Landes, B. G.; Quintana, J. P.; Ree, M. *Macromolecules* 2005, 38 (22), 8991-8995.

[75] Heo, K.; Jin, K. S.; Yoon, J.; Jin, S.; Oh, W.; Ree, M. *J. Phys. Chem. B* 2006, 110, 15887-15895.

[76] Nguyen, C.; Hawker, C. J.; Miller, R. D.; Huang, E.; Hedrick, J. L.; Gauderon, R.; Hilborn, J. G. *Macromolecules* 2000, 33 (11), 4281-4284.

[77] Mecerreyes, D.; Huang, E.; Magbitang, T.; Volksen, W.; Hawker, C. J.; Lee, V. Y.; Miller, R. D.; Hedrick, J. L. *High Perform. Polym.* 2001, 13 (2), s11-s19.

[78] Mecerreyes, D.; Atthoff, B.; Boduch, K. A.; Trollsas, M.; Hedrick, J. L. *Macromolecules* 1999, 32 (16), 5181-5182.

[79] Plummer, C. J. G.; Garamszegi, L.; Nguyen, T. Q.; Rodlert, M.; Manson, J. A. E. *J. Mater. Sci.* 2002, 37 (22), 4819-4829.

[80] Kim, J. S.; Kim, H. C.; Lee, B.; Ree, M. *Polymer* 2005, 46 (18), 7394-7402.

[81] Bergbreiter, D. E. *Angew. Chem. Int. Ed.* 1999, 38 (19), 2870-2872.

[82] Monteiro, M. J.; Bussels, R.; Wilkinson, T. S. *J. Polym. Sci.: Part A: Polym. Chem.* 2001, 39 (16), 2813-2820.

[83] Park, M. K.; Xia, C.; Advincula, R. C.; Schutz, P.; Caruso, F. *Langmuir* 2001, 17 (24), 7670-7674.

[84] Demers, L. M.; Park, S. J.; Andrew Taton, T.; Li, Z.; Mirkin, C. A. *Angew. Chem. Int. Ed.* 2001, 40 (16), 3071-3073.

[85] Thurmond Ii, K. B.; Kowalewski, T.; Wooley, K. L. *J. Am. Chem. Soc.* 1997, 119 (28), 6656-6665.

[86] Zhang, Q.; Remsen, E. E.; Wooley, K. L. *J. Am. Chem. Soc.* 2000, 122 (15), 3642-3651.
[87] Harth, E.; Van Horn, B.; Lee, V. Y.; Germack, D. S.; Gonzales, C. P.; Miller, R. D.; Hawker, C. J. *J. Am. Chem. Soc.* 2002, 124 (29), 8653-8660.
[88] Mecerreyes, D.; Lee, V.; Hawker, C. J.; Hedrick, J. L.; Wursch, A.; Volksen, W.; Magbitang, T.; Huang, E.; Miller, R. D. *Adv. Mater.* 2001, 13 (3), 204-208.
[89] Connor, E. F.; Sundberg, L. K.; Kim, H. C.; Cornelissen, J. J.; Magbitang, T.; Rice, P. M.; Lee, V. Y.; Hawker, C. J.; Volksen, W.; Hedrick, J. L.; Miller, R. D. *Angew. Chem. Int. Ed.* 2003, 42 (32), 3785-3788.
[90] Hong-ji, C.; Meng, F. *Macromolecules* 2007, 40, 2079-2085.
[91] Hwang, Y. T.; Yoon, J.; Oh, W.; Lee, B.; Ree, M. *unpublished results*.
[92] Carter, K. R.; Dawson, D. J.; Dipietro, R. A.; Hawker, C. J.; Hedrick, J. L.; Miller, R. D.; Yoon, D. Y. *US Pat. 5895263* 1999.
[93] Huang, Q. R.; Volksen, W.; Huang, E.; Toney, M.; Frank, C. W.; Miller, R. D. *Chem. Mater.* 2002, 14 (9), 3676-3685.
[94] Kim, H. C.; Wilds, J. B.; Kreller, C. R.; Volksen, W.; Brock, P. J.; Lee, V. Y.; Magbitang, T.; Hedrick, J. L.; Hawker, C. J.; Miller, R. D. *Adv. Mater.* 2002, 14 (22), 1637-1639.
[95] Kim, H. C.; Volksen, W.; Miller, R. D.; Huang, E.; Yang, G.; Briber, R. M.; Shin, K.; Satija, S. K. *Chem. Mater.* 2003, 15 (3), 609-611.
[96] Huang, Q. R.; Kim, H. C.; Huang, E.; Mecerreyes, D.; Hedrick, J. L.; Volksen, W.; Frank, C. W.; Miller, R. D. *Macromolecules* 2003, 36 (20), 7661-7671.
[97] Chang, Y.; Chen, C. Y. I.; Chen, W. C. *J. Polym. Sci., Part B: Polym. Phys.* 2004, 42 (24), 4466-4477.
[98] Yang, C. C.; Wu, P. T.; Chen, W. C.; Chen, H. L. *Polymer* 2004, 45 (16), 5691-5702.
[99] Yang, S.; Mirau, P. A.; Pai, C. S.; Naiamasu, O.; Reichmanis, E.; Lin, E. K.; Lee, H. J.; Gidley, D. W.; Sun, J. *Chem. Mater.* 2001, 13 (9), 2762-2764.
[100] Xu, J.; Moxom, J.; Yang, S.; Suzuki, R.; Ohdaira, T. *Chem. Phys. Let.* 2002, 364 (3-4), 309-313.

[101] Yang, S.; Mirau, P. A.; Pai, C. S.; Nalamasu, O.; Reichmanis, E.; Pai, J. C.; Obeng, Y. S.; Seputro, J.; Lin, E. K.; Lee, H. J.; Sun, J.; Gidley, D. W. *Chem. Mater.* 2002, 14 (1), 369-374.
[102] Polarz, S.; Smarsly, B.; Bronstein, L.; Antonietti, M. *Angew. Chem.* 2001, 113 (23), 4549-4553.
[103] Yim, J. H.; Lyu, Y. Y.; Jeong, H. D.; Song, S. A.; Hwang, I. S.; Hyeon-Lee, J.; Mah, S. K.; Chang, S.; Park, J. G.; Hu, Y. F.; Sun, J. N.; Gidley, D. W. *Adv. Func. Mater.* 2003, 13 (5), 382-386.
[104] Yim, J. H.; Seon, J. B.; Jeong, H. D.; Pu, L. S.; Baklanov, M. R.; Gidley, D. W. *Adv. Func. Mater.* 2004, 14 (3), 277-282.
[105] Yim, J. H.; Baklanov, M. R.; Gidley, D. W.; Peng, H.; Jeong, H. D.; Pu, L. S. *J. Phys. Chem. B* 2004, 108 (26), 8953-8959.
[106] Shin, J. J.; Park, S. J.; Min, S. K.; Rhee, H. W.; Moon, B.; Yoon, D. Y. *Mole. Cryst. Liq. Cryst.* 2006, 445, 167-175.
[107] Chung, K.; Moyer, E. S.; Spaulding, M. *US Pat. 6231989* 2001.
[108] Zhong, B.; Spaulding, M.; Albaugh, J.; Moyer, E. *Polym. Mater. Sci. Eng.* 2002, 87, 440-441.
[109] Oh, W. Ph.D. Thesis, Pohang University of Science and Technology, Rep. Korea, 2003.
[110] Cha, B. J.; Kim, S.; Char, K.; Lee, J. K.; Yoon, D. Y.; Rhee, H. W. *Chem. Mater.* 2006, 18, 378-385.
[111] Hyeon-Lee, J.; Kim, W. C.; Min, S. K.; Ree, H. W.; Yoon, D. Y. *Macro. Mater. Eng.* 2003, 288 (5), 455-461.
[112] Ro, H. W.; Kim, K. J.; Theato, P.; Gidley, D. W.; Yoon, D. Y. *Macromolecules* 2005, 38 (3), 1031-1034.
[113] Plummer, C. J. G.; Hilborn, J. G.; Hedrick, J. L. *Polymer* 1995, 36 (12), 2485-2489.
[114] Cha, H. J.; Hedrick, J.; DiPietro, R. A.; Blume, T.; Beyers, R.; Yoon, D. Y. *Appl.Phys. Lett.* 1996, 68 (14), 1930-1932.
[115] Carter, K. R.; Di Pietro, R.; Sanchez, M. I.; Russell, T. P.; Lakshmanan, P.; McGrath, J. E. *Chem. Mater.* 1997, 9, 105-118.
[116] Do, J. S.; Zhu, B.; Han, S. H.; Nah, C.; Lee, M. H. *Polym. Int.* 2004, 53 (8), 1040-1046.
[117] Charlier, Y.; Hedrick, J. L.; Russell, T. P.; Jonas, A.; Volksen, W. *Polymer* 1995, 36 (5), 987-1002.
[118] Hedrick, J. L.; Hawker, C. J.; DiPietro, R.; Jerome, R.; Charlier, Y. *Polymer* 1995, 36 (25), 4855-4866.

[119] Hedrick, J. L.; Carter, K.; Richter, R.; Miller, R. D.; Russell, T. P.; Flores, V.; Mecerreyes, D.; Dubois, P.; Jerome, R. *Chem. Mater.* 1998, 10, 39-49.
[120] Hedrick, J.; Labadie, J.; Russell, T.; Hofer, D.; Wakharker, V. *Polymer* 1993, 34, (22), 4717-4726.
[121] Fu, G. D.; Zong, B. Y.; Kang, E. T.; Neoh, K. G.; Lin, C. C.; Liaw, D. J. *Indus. Eng. Chem. Res.* 2004, 43 (21), 6723-6730.
[122] Fu, G. D.; Wang, W. C.; Li, S.; Kang, E. T.; Neoh, K. G.; Tseng, W. T.; Liaw, D. J. *J. Mater. Chem* 2003, 13 (9), 2150-2156.
[123] Chen, Y.; Wang, W.; Yu, W.; Yuan, Z.; Kang, E. T.; Neoh, K. G.; Krauter, B.; Greiner, A. *Adv. Func. Mater.* 2004, 14 (5), 471-478.
[124] Carter, K. R.; DiPietro, R. A.; Sanchez, M. I.; Swanson, S. A. *Chem. Mater.* 2001, 13 (1), 213-221.
[125] Chen, Y. W.; Wang, W. C.; Yu, W. H.; Kang, E. T.; Neoh, K. G.; Vora, M. H.; Ong, C. K.; Chen, L. F. *J. Mater. Chem.* 2004, 14 (9), 1406-1412.
[126] Wang, W. C.; Vora, R. H.; Kang, E. T.; Neoh, K. G.; Ong, C. K.; Chen, L. F. *Adv. Mater.* 2004, 16 (1), 54-57.
[127] Oh, W.; Shin, T. J.; Ree, M.; Jin, M. Y.; Char, K. *Macromol. Chem. Phys.* 2002, 20 (5-6), 801-811.
[128] Rajagopalan, T.; Lahlouh, B.; Lubguban, J. A.; Biswas, N.; Gangopadhyay, S.; Sun, J.; Huang, D. H.; Simon, S. L.; Mallikarjunan, A.; Kim, H. C.; Volksen, W.; Toney, M. F.; Huang, E.; Rice, P. M.; Delenia, E.; Miller, R. D. *Appl. Phys. Lett.* 2003, 82 (24), 4328-4330.
[129] Heo, K.; Yoon, J.; Jin, K. S.; Jin, S.; Ree, M. *IEE Proc. Bionanotech.* 2006, 153, 121-128.
[130] Hsu, C.-H.; Jeng, U.-S.; Lee, H.-Y.; Huang, C.-M.; Liang, K. S.; Windover, D.; Lu, T.-M.; Jin, C. *Thin Solid Films* 2005, 472, 323–327.
[131] Jousseaume, V.; Rolland, G.; Babonneau, D.; Simon, J.-P. *Appl. Surf. Sci.* 2007, 254, 473-479.
[132] Omote, K.; Ito, Y.; Kawamura, S. *Appl. Phys. Lett.* 2003, 82, 544-546.
[133] Lee, B.; Park, I.; Yoon, J.; Park, S.; Kim, J.; Kim, K. W.; Chang, T.; Ree, M. *Macromolecules* 2005, 38 (10), 4311-4323.
[134] Heo, K.; Jin, K. S.; Oh, W.; Yoon, J.; Jin, S.; Ree, M. *J. Phys. Chem. B* 2006, 110, 15887-15895.

[135] Tolan, M., *X-Ray Scattering form Soft-Matter Thin Films.* Springer: New York, 1998.
[136] Holy, V.; Pietsch, U.; Baumbach, T., *High-Resolution X-ray Scattering from Thin Films and Multilayers.* Springer: New York, 1999.
[137] Lazzari, R. *J. Appl. Crystallog.* 2002, 35 (4), 406-421.
[138] Heo, K.; Park, S.-G.; Yoon, J.; Jin, K. S.; Jin, S.; Rhee, S.-W.; Ree, M. *J. Phys. Chem. C* 2007, 111, 10848-10854.
[139] Heo, K.; Yoon, J.; Jin, S.; Kim, J.; Kim, K.-W.; Shin, T. J.; Chung, B.; Chang, T.; Ree, M. *J. Appl. Crystallogr.* 2008, 41, 281-291.
[140] Jin, S.; Yoon, J.; Heo, K.; Park, H.-W.; Shin, T. J.; Chang, T.; Ree, M. *J. Appl. Crystallogr.* 2007, 40, 950-958.
[141] Yoon, J.; Choi, S. C.; Jin, S.; Jin, K. S.; Heo, K.; Ree, M. *J. Appl. Crystallogr.* 2007, 40, s669-s674.
[142] Yoon, J.; Yang, S. Y.; Heo, K.; Lee, B.; Joo, W.; Kim, J. K.; Ree, M. *J. Appl. Crystallogr.* 2007, 40, 305-312.
[143] Harmer, M. A.; Farneth, W. E.; Sun, Q. *J. Am. Chem. Soc.* 1996, 118 (33), 7708-7715.
[144] Kim, S.; Toivola, Y.; Cook, R. F.; Char, K.; Chu, S.-H.; Lee, J.-K.; Yoon, D. Y.; Rhee, H.-W. *J. Electrochem. Soc.* 2004, 151 (3), F37-F44.
[145] Pedersen, J. S. *J. Appl. Crystallogr.* 1994, 27, 595-608.
[146] Lin, E. K.; Lee, H. J.; Lynn, G. W.; Wu, W. L.; O'Neill, M. L. *Appl. Phys. Lett.* 2002, 81 (4), 607.
[147] Lee, H.-J.; Lin, E. K.; Bauer, B. J.; Wu, W.-l.; Hwang, B. K.; Gray, W. D. *Appl. Phys. Lett.* 2003, 82, 1084-1086.
[148] Lee, H. J.; Lin, E. K.; Wang, H.; Wu, W. L.; Chen, W.; Moyer, E. S. *Chem. Mater.* 2002, 14 (4), 1845-1852.
[149] Huang, Q. R.; Volksen, W.; Hunag, E.; Toney, M.; Frank, C. W.; Miller, R. D. *Chem. Mater.* 2002, 14, 3676-3685.
[150] Hedden, R. C.; Lee, H. J.; Soles, C. L.; Bauer, B. J. *Langmuir* 2004, 20 (16), 6658-6667.
[151] Wu, W. L.; Wallace, W. E.; Lin, E. K.; Lynn, G. W.; Glinka, C. J.; Ryan, E. T.; Ho, H. M. *J. Appl. Phys.* 2000, 87 (3), 1193-1200.
[152] Yang, S.; Mirau, P. A.; Pai, C. S.; Naiamasu, O.; Reichmanis, E.; Lin, E. K.; Lee, H. J.; Gidley, D. W.; Sun, J. *Chem. Mater.* 2001, 13 (9), 2762-2764.

[153] Huang, E.; Toney, M. F.; Volksen, W.; Mecerreyes, D.; Brock, P.; Kim, H. C.; Hawker, C. J.; Hedrick, J. L.; Lee, V. Y.; Magbitang, T.; Miller, R. D.; Lurio, L. B. *Appl. Phys. Lett.* 2002, 81 (12), 2232-2234.
[154] Chamard, V.; Bastie, P.; Le Bolloch, D.; Dolino, G.; Elkaim, E.; Ferrero, C.; Lauriat, J. P.; Rieutord, F.; Thiaudiere, D. *Phys. Rev. B* 2001, 64 (24), 2454161-2454164.
[155] Hedden, R. C.; Waldfried, C.; Lee, H.-J.; Escorcia, O. *J. Electrochem. Soc.* 2004, 151 (8), F178-F181.
[156] Lee, T. J.; Byun, G.-S.; Jin, K. S.; Heo, K.; Kim, G.; Kim, S. Y.; Cho, I.; Ree, M. *J. Appl. Crystallogr.* 2007, 40, s620-s625
[157] Oh, W.; Hwang, Y.; Shin, T. J.; Lee, B.; Kim, J.-S.; Yoon, J.; Brennan, S.; Mehta, A.; Ree, M. *J. Appl. Crystallogr.* 2007, 40, s626-s630.
[158] Hatton, B. D.; Landskron, K.; Whitnall, W.; Perovic, D. D.; Ozin, G. A. *Adv. Funct. Mater.* 2005, 15 (5), 823-829.
[159] Mori, H.; Lanzendorfer, M. G.; Muller, A. H. E. *Macromolecules* 2004, 37, 5228-5238.
[160] Toivola, Y.; Kim, S.; Cook, R. F.; Char, K.; Lee, J.-K.; Yoon, D. Y.; Rhee, H.-W.; Kim, S. Y.; Jin, M. Y. *J. Electrochem. Soc.* 2004, 151 (3), F45-F53.
[161] Silverstein, M. S.; Shach-Caplan, M.; Bauer, B. J.; Hedden, R. C.; Lee, H.-J.; Landes, B. G. *Macromolecules* 2005, 38, 4301-4310.
[162] Li, G.; Zhou, H.; Honma, I. *Nat. Mater.* 2004, 3, 65-72.
[163] Gregg, S. J.; Sing, K. S. W. *Adsorption, Surface Area and Porosity* 1982.
[164] Baklanov, M. R.; Dultsev, F. N.; Repinsky, S. M. *Poverkhnost* 1988, 11, 145-146.
[165] Baklanov, M. R.; Vasilyeva, L. L.; Gavrilova, T. A.; Dultsev, F. N.; Mogilnikov, K. P.; Nenasheva, L. A. *Thin Solid Films* 1989, 171 (1), 43-52.
[166] Ramsay, J. D. F. *MRS Bulletin* 1999, 24 (3), 36-40.
[167] Baklanov, M. R.; Mogilnikov, K. P.; Polovinkin, V. G.; Dultsev, F. N. *J. Vac. Sci. Technol. B* 2000, 18 (3), 1385-1391.
[168] Dultsev, F. N.; Baklanov, M. R. *Electrochem. Solid-State Lett.* 1999, 2 (2-4), 192-194.
[169] Baklanov, M. R.; Mogilnikov, K. P. *Microeletron. Eng* 2002, 64 (1-4), 335-349.

[170] Othman, M. T.; Lubguban, J. A.; Lubguban, A. A.; Gangopadhyay, S.; Miller, R. D.; Volksen, W.; Kim, H.-C. *J. Appl. Phys.* 2006, 99, 083503-1-7.
[171] Shamiryan, D.; Baklanov, M. R.; Maex, K. *J. Vac. Sci. Technol. B* 2003, 21 (1), 220-226.
[172] Petkov, M. P.; Weber, M. H.; Lynn, K. G.; Rodbell, K. P.; Cohen, S. A. *Appl. Phys. Lett.* 1999, 74, 2146-2148.
[173] Gidley, D. W.; Frieze, W. E.; Dull, T. L.; Sun, J.; Yee, A. F.; Nguyen, C. V.; Yoon, D. Y. *Appl. Phys. Lett.* 2000, 76, 1282-1284.
[174] Petkov, M. P.; Weber, M. H.; Lynn, K. G.; Rodbell, K. P. *Appl. Phys. Lett.* 2000, 77, 2470-2472.
[175] Sun, J. N.; Hu, Y. F.; Frieze, W. E.; Gidley, D. W. *Rad. Phys. Chem.* 2003, 68, 345-349.
[176] Dull, T. L.; Frieze, W. E.; Gidley, D. W.; Sun, J. N.; Yee, A. F. *J. Phys. Chem. B* 2001, 105 (20), 4657-4662.
[177] Sun, J. N.; Hu, Y.; Frieze, W. E.; Chen, W.; Gidley, D. W. *J. Electrochem. Soc.* 2003, 150 (5), F97-F101.
[178] Sun, J. N.; Gidley, D. W.; Dull, T. L.; Frieze, W. E.; Yee, A. F.; Ryan, E. T.; Lin, S.; Wetzel, J. *J. Appl. Phys.* 2001, 89 (9), 5138-5144.
[179] Gidley, D. W.; Frieze, W. E.; Dull, T. L.; Yee, A. F.; Ryan, E. T.; Ho, H. M. *Phys. Rev. B* 1999, 60 (8), R5157-R5160.
[180] Sun, J. N.; Gidley, D. W.; Hu, Y.; Frieze, W. E.; Ryan, E. T. *Appl. Phys. Lett.* 2002, 81 (8), 1447.
[181] Peng, H. G.; Frieze, W. E.; Vallery, R. S.; Gidley, D. W.; Moore, D. L.; Carter, R. J. *Appl. Phys. Lett.* 2005, 86 (12), 1-3.
[182] Tian, D.; Blacher, S.; Pirard, J. P.; Jerome, R. *Langmuir* 1998, 14 (7), 1905-1910.
[183] Grill, A.; Patel, V.; Rodbell, K. P.; Huang, E.; Baklanov, M. R.; Mogilnikov, K. P.; Toney, M.; Kim, H. C. *J. Appl. Phys.* 2003, 94 (5), 3427-3435.
[184] Yu, S.; Wong, T. K. S.; Hu, X.; Pita, K. *J. Electrochem. Soc.* 2003, 150, (5), F116-F121

# INDEX

## A

acid, 2, 15
acrylate, 7, 8, 9
acrylic acid, 21
adamantane, 25
adaptation, 43
adhesion, 2
adsorption, 29, 43, 44, 46, 49
aerogels, 4
AFM, 21, 29, 42, 46
aggregates, 12, 38, 39
aggregation, 7, 9, 11, 12, 14, 17, 18, 22, 37, 47, 48
alcohol, 39
aliphatic compounds, 21
aliphatic polymers, 12, 14
amines, 23
ammonia, 24
ammonium, 2
annealing, 2
annihilation, 21, 29, 44
atomic force, 29
atoms, 49
azimuthal angle, 31

## B

barriers, 2
beams, 40, 42, 46
behavior, 2, 11
benzene, 15
blend films, 10, 12, 37
blends, 20
blocks, 18, 26, 27
bonding, 19, 20
branched polymers, 10
branching, 5
burn, 6

## C

calibration, 45
candidates, 6, 12
capillary, 4, 27, 48
casting, 6, 24
catalyst, 24
catalytic activity, 9
cell, 20
chemical deposition, 3
chemical interaction, 10
chemical properties, 29, 49
chemical vapor deposition, 2, 3
cleavage, 25

collisions, 44
competition, 38
components, 5, 6, 20, 31, 48
composition, 20
conductor, vii, 1, 44
control, vii, 48
cooling, 37
copolymers, 6, 15, 18, 19, 20, 24, 25
copper, 1, 44
correlation, 40
correlations, 30, 46
coupling, 1, 20
covalent bond, 25
crack, 43
crystalline, 14
curing, 6, 12, 23, 24, 25, 26, 38, 39
curing reactions, 6, 12, 23
cycles, 4, 27, 48

## D

defects, 3, 29
degradation, 6, 17, 20, 27
density, 1, 30, 36, 37, 39, 40
deposition, 2
deposits, 2
derivatives, 4, 10
desorption, 43
dielectric constant, vii, 32
dielectrics, vii, 2, 3, 4, 6, 10, 12, 14, 15, 17, 19, 20, 21, 22, 24, 25, 26, 27, 29, 47, 48, 49
diffusion, 2, 44
dispersion, 17, 33, 48
distribution, vii, 3, 9, 12, 13, 20, 29, 30, 34, 35, 36, 37, 38, 40, 43, 44, 45, 46, 47, 49
distribution function, 35
drying, 4, 6, 24, 47

## E

Egypt, 53

electric circuits, 1
electron, 8, 14, 15, 29, 30, 34, 36, 37, 39, 40, 42, 44, 45
electron microscopy, 8, 14, 15, 29, 42
electrons, 44, 45
electroplating, 2
encapsulation, 5
energy, 41
ester, 10
etching, 2
ethanol, 12
ethers, 3
ethyl alcohol, 39
ethylene glycol, 27
ethylene oxide, 18, 19, 20

## F

fabrication, 3, 4, 6, 16, 27, 45, 47, 49
field emission scanning electron microscopy, 29
film formation, 6, 12, 14, 47
film thickness, 29, 31, 33, 34, 36, 40, 49
films, 2, 4, 7, 9, 10, 11, 12, 13, 14, 15, 16, 17, 18, 20, 21, 22, 23, 24, 25, 27, 30, 31, 32, 34, 35, 36, 39, 43, 46, 47
fluoropolymers, 3
free radicals, 45
free volume, 44, 45

## G

gel, 24, 25
generation, 12, 23, 30, 37, 49
genre, 43
glass transition, 26
glass transition temperature, 26
glucose, 21
grazing, 2, 9, 13, 29, 30, 33
groups, 5, 7, 9, 10, 12, 14, 19, 20, 23, 24, 25, 37
growth, 38

# Index

## H

heat, 12, 17, 47
heating, 23, 37, 38, 39
heptane, 43
homopolymers, 19
HRTEM, 29, 42, 46
hybrid, 25, 26, 27, 39
hybridization, 38
hydrogen, 19, 20
hydrosilylation, 23, 25
hydroxide, 2
hydroxyl, 10, 12, 22
hyperbranched polymers, 14, 15

## I

image, 8, 42
images, 42
imprinting, vii, 2, 4, 6, 10, 37
incidence, 2, 7, 9, 13, 29, 30, 31, 33
industry, 1, 43, 47, 49
integrated circuits, vii, 1, 44
integration, 1, 30, 49
interaction, 46
interactions, 20, 21
interface, 16, 20, 32, 35, 40
interfacial adhesion, 3, 5

## K

Korea, 49, 51, 53, 54, 57

## L

lactic acid, 19
lifetime, 21, 29, 44, 45
limitation, 15
linear polymers, 10
linkage, 21

## M

macromolecules, 17
matrix, 5, 7, 8, 12, 14, 16, 17, 18, 20, 23, 24, 35, 36, 38, 39, 40, 48
mechanical properties, 3, 5
media, 40
metals, 2
methyl methacrylate, 15, 17, 19
microelectronics, 1, 49
microscopy, 21, 29, 42, 46, 49
Ministry of Education, 49
MMA, 19, 20
model, 10, 34, 35, 40, 46
modulus, 25
moisture, 2, 6, 32
molecular mobility, 4, 27, 48
molecular weight, 6, 17, 25
molecules, 6, 7, 11, 12, 14, 19, 23, 24, 36, 37, 44
morphology, 40
movement, 38

## N

nanometer, 5, 10, 14, 17, 22, 29, 48
nanometer scale, 5, 48
nanoparticles, 4, 5, 17, 19, 47
nanotechnology, 17
network, 6, 24, 48
network polymers, 48
nitrogen, 6, 23, 24, 43, 44
norbornene, 18

## O

order, 3, 29, 48
organic dielectrics, 47
organic polymers, 4, 48
organic solvents, 24

## P

PAA, 21
parameters, 36, 37, 40
particles, 2, 4, 17, 37, 47
plasma, 2, 3
PMMA, 10, 25, 26
poly(methyl methacrylate), 10, 25
polycondensation, 6, 9, 15, 19, 23
polyesters, 19
polyether, 5
polyimides, 3, 4, 5, 26, 27, 45, 48
polymer, 4, 11, 12, 14, 15, 16, 17, 18, 19, 26, 27
polymerization, 15
polymers, 3, 10, 15, 18, 19, 26
polystyrene, 26
porosity, 12, 17, 20, 21, 23, 24, 25, 30, 36, 37, 40, 43, 48
porous materials, 40
positron, 29, 44, 45, 46
positrons, 45
precipitation, 18, 20
pressure, 4, 23, 27, 48
probe, 43, 46
production, vii, 1, 4, 17, 47, 49
propylene, 5, 19, 20
PTFE, 3
purity, 2

## Q

quartz, 43

## R

radiation, 11, 29, 40, 41, 46
radical copolymerization, 19
radius, 7, 9, 13, 14, 15, 16, 17, 34, 35, 36, 37, 39, 41
radius of gyration, 7
range, 2, 6, 7, 9, 10, 13, 14, 15, 17, 20, 21, 22, 23, 24, 25, 36, 38, 47
reflection, 32, 41
reflectivity, 20, 29, 40, 41
refractive index, 4, 13, 21, 22, 26, 33, 43
refractive indices, 9, 11, 16, 17, 20
region, 39, 41
reliability, 5, 44
resolution, 29, 41, 42
rings, 20
room temperature, 19, 23, 37, 43
roughness, 31, 40, 43

## S

scaling, 1
scanning electron microscopy, 29
scattering, 2, 7, 9, 13, 20, 21, 29, 30, 31, 32, 33, 34, 35, 36, 37, 39, 40, 41, 46
scattering patterns, 37
seed, 44
segregation, 12
semiconductor, 47, 49
sensing, 43
sensitivity, 41, 45
separation, 6, 11, 19, 20, 38
severe stress, 43
shape, 5, 6, 10, 11, 12, 14, 17, 30, 32, 37, 42
silane, 23
silica, 4
silicon, 2, 3, 20, 31, 32, 34, 36, 40, 41, 43, 44
silver, 1
sintering, 4
sol-gel, 4, 19, 21, 25
solubility, 5, 16, 17
solvents, 9, 16, 24
space, 7, 12
spectroscopy, 21, 29, 44, 46
spin, vii, 2, 4, 6, 19, 47
stability, 4, 6, 47, 48
STM, 29, 42, 46
strength, 32, 46
stress, 2
structural characteristics, 12

styrene, 17, 20, 21
substrates, 18, 20, 31
swelling, 44

## T

TEM, 8, 15, 18, 21, 22, 29, 42, 45, 46
temperature, 4, 6, 23, 26, 37, 38, 39
TEOS, 44
tetraethoxysilane, 44
thermal decomposition, 6, 7, 16, 25
thermal degradation, 6, 10, 14, 19, 22, 25, 39, 42, 47
thermal evaporation, 23
thermal expansion, 40
thermal stability, 6, 32
thermal treatment, 6, 12, 20, 25
thermolysis, 27
thin films, 2, 3, 4, 7, 16, 29, 30, 32, 37, 40, 41, 43, 46, 47, 49
tin, 30
toluene, 43

transition, 40
transmission, 30, 32, 40
transmission electron microscopy, 29
tunneling, 29

## U

uniform, vii, 3, 17, 29

## V

vacuum, 3, 23, 31, 38, 42, 44, 45
vapor, 2, 23, 24, 43
vector, 31, 33
viscosity, 5

## W

water absorption, 19
wave vector, 33